Introductory Elements of Analysis and Design in Chemical Engineering

Introductory Elements of Analysis and Design in Chemical Engineering introduces readers to how chemical engineers think. It explains the application of analytical methods to phenomena important in chemical engineering and teaches analytical skills in the context of engineering design. A principle goal is to help readers reinforce their understanding of mathematics (especially calculus) and science as they are introduced to engineering thinking.

Key Features:

- Emphasizes basic principles, methods, and problem solving at an elementary level.
- Presents concepts in calculus, chemistry, and physics and methods of analysis on the basis of experiment and observation.
- Connects experimental results to mathematical representations.
- Provides numerous illustrative examples and builds on them to introduce processing and process flow diagrams and to place chemical engineering in an historical context.
- Includes problems at the end of each chapter.

Aimed at readers beginning their studies in chemical engineering, this textbook offers an approachable introduction to the principles of analysis and design in chemical engineering to help readers learn to think quantitatively and with a foundation of chemical engineering concepts.

Introductory Elements of Analysis and Design in Chemical Engineering

Bruce C. Gates

With a chapter coauthored by Robert L. Powell

CRC Press
Taylor & Francis Group
Boca Raton London New York

CRC Press is an imprint of the
Taylor & Francis Group, an **informa** business

First edition published 2024
by CRC Press
2385 Executive Center Drive, Suite 320, Boca Raton, FL 33431

and by CRC Press
4 Park Square, Milton Park, Abingdon, Oxon, OX14 4RN

CRC Press is an imprint of Taylor & Francis Group, LLC

ISBN: 978-1-032-55288-0 (hbk)
ISBN: 978-1-032-55283-5 (pbk)
ISBN: 978-1-003-42994-4 (ebk)

DOI: 10.1201/9781003429944

Typeset in Times
by codeMantra

Dedication

_____

For Lexi, Lilja, and Micha

Contents

Preface

Engineers use mathematics as a language to represent real phenomena. The procedures for identifying, unraveling, and representing the phenomena are called analysis, and the representations take the form of equations that are called models. Models are used for prediction of phenomena, planning of experiments, and design of equipment and processes. Good engineering models are simplified representations of reality—approximations that should be as simple as possible while representing the essence of the phenomena. They must be mathematically correct and should be consistent with scientific principles. These models guide engineering design, which benefits from the insights that underlie them—combined with creativity, judgment, and experience.

Most engineering courses are taught on the premise that students learn analysis by osmosis and example and that they learn design on the basis of skills in analysis. Skills in analysis are usually not taught explicitly in mathematics or science courses. In our experience, engineering students benefit from an early and focused introduction to analysis placed in the context of design. This book emerged from our experience starting about 15 years ago teaching an introduction to the ideas of analysis and design to students at the beginning of their university study.

Our goal is to present an introduction to how engineers think by application of the methods of analysis to phenomena important in chemical engineering. We assume that students will be learning calculus, chemistry, and possibly physics as they are learning from this book. The concepts and methods of analysis are introduced on the basis of experiment and observation. Students who may still be coming to grips with concepts of calculus (such as limits) are guided in understanding engineering analysis with this physical approach. One of our principle goals is to help students reinforce their understanding of mathematics (especially calculus) and science as they are introduced to engineering thinking. We have restricted our topics to keep the mathematics at the level of introductory calculus, and therefore do not come close to presenting a complete picture of chemical engineering.

We start in the laboratory with a simple experiment, the draining of water from a tank through a hole in its base. On the basis of measurements of the liquid height as a function of time, we develop an equation to represent the rate of water flow. The concept of rate is central to this and further developments in this book, linking engineering analysis and calculus. We use results of the analyses to design systems with combinations of tanks and constant-head tanks. We have attempted to make the topics of design broad and intriguing to students, because many of them choose engineering because they like design.

Another example is a burning candle; this is easy to observe but complex in the phenomena that occur, and we do not analyze it in depth, but rather use it to link beginning chemistry to our topic, and to reinforce the idea of a rate. We ask how fast the candle burns and how that can be measured—and how it works: what phases (gas, liquid, and solid) are present, and what individual processes take place as the candle burns (chemical reaction, phase changes, fluid flow, and heat transport).

Further examples involve measuring rates of a chemical reaction and converting acorns into food. The latter involves multiple steps such as extraction that are commonly encountered in chemical processing and food processing. We build from such examples to introduce the ideas of processing and process flow diagrams, and to begin to place chemical engineering in an historical context, emphasizing the value of experimentation as part of the learning process.

Thus, flow, transport, separation, and chemical change are considered quantitatively in increasingly complex combinations in processes that, like candle burning, often operate on a continuous basis. Fundamental principles that underlie the presentation are conservation of mass and conservation of energy, and we illustrate the use of mass balance and energy balance equations as valuable starting points for analysis. We also introduce chemical change and chemical reaction rates, linking equations describing them with equations describing ideal chemical reactors as a basis for design of reactors. And we illustrate reactors for direct determination and visualization of reaction rates.

In learning how to determine equations to represent observations, we initially use physical and mathematical reasoning to find out what equations work well, and then go beyond this empirical approach to be guided by dimensional analysis and fundamental scientific principles including conservation of mass, conservation of energy, and the fundamentals of reaction kinetics.

Beyond observing and unraveling phenomena and representing them mathematically and schematically in process flow diagrams, we use models to predict performance and design simple equipment and processes, some linked, for example, to draining tanks. We attempt to bring more and more examples of design and more integration of topics into this book as we get deeper into the subject, and we attempt to integrate ideas and examples introduced in early chapters into later chapters.

The development presented here relies on visualization based on experiment, calculations (with an emphasis on graphical methods), and elementary mathematical operations at the level of beginning calculus. The experimentation in our view is an essential part of the learning experience for students, and we believe that this book will be more effective as a learning tool if experimentation is part of the student experience. Many of the examples considered here involve experiments that are relatively safe and equipment that is inexpensive and does not require much space. Videos of some of the experiments will be available online.

This book thus emphasizes basic principles and methods—and problem solving to help students learn. We provide examples throughout the text and problems at the end of each chapter. We encourage group activity in problem solving to help students develop their networking and learning skills and appreciate how others learn.

Computers and software are not essential at this early stage of the curriculum, and to maintain the focus on principles and methods, we resist their charm and efficiency (which greatly attract students) and defer implementation of those tools. We recommend subsequent learning for students with the more advanced book *Introduction to Chemical Engineering Analysis Using Mathematica* by Henry C. Foley (second edition, 2021, Academic Press).

In summary, our goal is to present an elementary introduction to the principles of analysis and design in chemical engineering to help students learn to think quantitatively and with a foundation of chemical engineering concepts. The subject is presented at a level that we consider appropriate not only for first-year university students in engineering but also for pre-engineering students. We hope that this book will help students develop the habit of figuring things out and creating designs rather than looking them up and repeating what is known.

We are grateful to our former students for helping us learn, and we respectfully appreciate the book *Introduction to Chemical Engineering Analysis* by the late T. W. F. Russell and M. M. Denn (Wiley, 1972), which we used as young teachers and which helped us to form our ideas.

Bruce C. Gates
Robert L. Powell
Davis, California, March 2023

Authors

Bruce C. Gates is Distinguished Professor Emeritus of Chemical Engineering at the University of California, Davis, where he has been a faculty member since 1992. He was a Research Engineer at Chevron Research Company and then the H. Rodney Sharp Professor of Chemical Engineering and Professor of Chemistry at the University of Delaware prior to joining the University of California.

Robert L. Powell is Distinguished Professor Emeritus of Chemical Engineering and of Food Science and Technology at the University of California, Davis, where he was a faculty member from 1984 to 2021. He was a faculty member in Chemical Engineering at Washington University in St. Louis from 1979–1984.

1 Introduction
Analysis, Design, and Chemical Engineering

ROADMAP

This chapter is a statement of the goals of this book, introducing terms that are essential for understanding the fundamental principles and methods used in chemical engineering: *analysis, synthesis, design, quantitative, qualitative, transient, equilibrium*, and *steady state*. Examples provide a context for explaining the meanings of these terms and starting us on the way to recognizing individual components that in combination constitute simple and easily visualized processes involving physical and chemical changes: (a) a draining tank, (b) tanks with inflow streams in addition to the outflow steam, and (c) a burning candle. This introductory chapter lays the foundation for the first, simple analysis to be done quantitatively (with equations and numbers), which appears in Chapter 2.

Goals of This Book

Engineers work to understand the physical world, and they use mathematics in the creative processes of converting their understanding into something useful, such as new or improved materials, instruments, devices, products, processes, or technologies—or intangible products such as advice, policies, or operating procedures. The purpose of this book is to help students get a good start into engineering by reinforcing their knowledge of mathematics and science to develop an understanding of systematic approaches to meeting engineering challenges—and to apply their understanding to design devices and processes. The thought processes of engineering analysis and design introduced here are not often taught in mathematics or science courses, or even in most beginning engineering courses such as stoichiometry, statics, or thermodynamics. Instead, students in those courses are often assumed to learn the thought processes by example, osmosis, and inference. One of our goals is to build a foundation that will help students succeed in subsequent engineering and science courses.

To begin, we introduce essential terms, concepts, and methods of engineering analysis. Because problem solving is essential to help students learn, this text is complemented with example problems and their solutions, and problem solving exercises at the end of each chapter. We build from simple examples and proceed to complex ones, illustrating concepts and methods. The exercises include design challenges so that students become aware of how analysis helps to facilitate design. The examples are intended to be easily visualized, because our foundation is meant to be physical

DOI: 10.1201/9781003429944-1

and to seem real rather than abstract or mathematical. A major goal is to help students develop a sense of physical reasoning that can be translated into successful analysis and understanding to guide the creative engineering process of design. Sound physical reasoning requires understanding of principles of science, and some are illustrated here with examples from beginning chemistry, physics, and biology.

In analysis, mathematics is used to represent the physical world. We observe the world experimentally and use mathematics as a language to represent it. The language must be mathematics, because only mathematics provides a concise representation of physical reality that is *quantitative*—that is, in numbers and equations. In using mathematics, we must play by its rules or expect to be wrong.

A Starting Example: A Draining Tank

We begin with a physical process that is simple and easy to picture, and we can easily go into a laboratory (or a kitchen) to observe it quantitatively. Water drains from a cylindrical tank through a circular opening in its base (called an orifice), represented schematically as follows:

We ask questions such as the following: How long does it take to drain the tank? How fast is the flow of water out of the tank when the tank is full, half full, and almost empty? How can we use the observations of draining of one tank to predict how long it takes to drain another tank, for example, one that is larger or one that has a larger orifice?

We make quantitative observations of the draining tank, determining the amount of water in it as a function of the time after the start of draining. We set out with the goal of representing our observations with an equation. We start with the simple approach of choosing an equation just on the basis of what works well to represent the observations (the data)—that is, what fits, what agrees well with the data. We call this an empirical approach; it involves approximations—simplifications of reality. Then we proceed to formulate more powerful and systematic approaches to the analysis, guided by a fundamental scientific principle called the conservation of mass—which states that mass is neither created nor destroyed—and guided by procedures that emerge from the requirement of mathematical consistency. Then we compare the equation with the data to see how well it works, and we use it to make predictions, such as for other tanks.

By working through this example, we see how mathematics works as a language that helps us to

- understand and analyze phenomena,
- systematize and generalize information, and
- make predictions and do design.

We proceed with the specifics of the example of the draining tank in the next chapter, but first need some definitions to explain where we are going.

Definitions to Start the Development

Two essential terms that appear in the preceding paragraphs require explanation, *analysis* and *design*; the corresponding verbs are *to analyze* and *to design*. First consider some paraphrased dictionary definitions of the term *to analyze*:

1. To separate or break up (any whole) into its (fundamental) parts to find out their nature, proportion, functional relationship, etc.
2. In chemistry, to separate (compounds or mixtures) into their constituent substances (elements) to determine the nature or proportion of the constituents.
3. In mathematics, to solve by means of equations.

These are different definitions for different specialized fields. The first definition is general and in line with what most of us understand by the word "analyze" in day-to-day conversation. The second definition is familiar to students of chemistry and the third to students of mathematics. As we work toward defining *engineering analysis* specifically, let us consider these meanings further and start applying them to the example of the draining tank.

The definitions of "to analyze" imply the idea of taking something apart, of parsing out its essential components. The taking apart connotes extraction of fundamental understanding, of capturing the essential individual ideas that, when combined, represent the whole. Thus, in analysis of physical phenomena, we seek to identify the important components or parts, discover their nature and function, and represent them by means of an equation or equations.

Another key term is *synthesis*. Let us consider some paraphrased dictionary definitions of the term *to synthesize*:

1. To bring components together into a whole.
2. To form by bringing together separate parts.
3. In chemistry, to form a (complex) chemical compound by combining two or more simpler compounds or elements.

Again, notice how there are general and specialized definitions and how the definition from chemistry is already familiar. Also notice that in a sense, analysis and synthesis are opposites. Analysis implies taking apart; synthesis implies putting together. And we know from chemistry that analysis and synthesis are complementary; if we want to synthesize a compound, we first need to know what it is composed of—we need to analyze it. In a broad sense, synthesis, forming a whole from its parts, requires that we know what the parts should be. So, in general, analysis precedes synthesis; it is part of the foundation of synthesis.

Let us consider analysis of the performance of the draining tank. We consider this to be a simple case, a kind of building block or component for our thought processes (Figure 1.1).

Next, consider a more complex physical situation, depicted in Figure 1.2.

Here, water flows out of the tank, as in Figure 1.1, and at the same time, water flows into the tank. One element of complexity has been added: the inflow stream.

Now, consider a still more complex situation, depicted in Figure 1.3; the various physical phenomena are listed to the right of the diagram.

The exercise we have just gone through is a kind of synthesis; we considered a simple physical situation, water draining from the tank, and added complications, new phenomena, as listed in Figure 1.3. We have built up the picture by considering

FIGURE 1.1   Water drains from a tank.

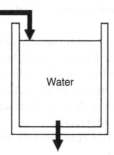

FIGURE 1.2   Water flows into a tank and simultaneously drains from it.

-Water flows into the tank
-Solid salt (NaCl) flows into the tank
-Salt and water are mixed in the tank
-Salt dissolves in water in the tank
-Water flows out of the tank
-Dissolved salt flows out of the tank
-Solid salt flows out of the tank

FIGURE 1.3   Tank with multiple components and multiple flows.

more than one stream and recognizing the processes of mixing and dissolution in addition to flow.

Now let us do something like the reverse of this synthesis, a primitive (and *only qualitative—that is, not quantitative*) analysis of the physical situation depicted in Figure 1.3. First, recognize that there is one outlet stream, which we might describe with the same terms used for the outlet stream in the simplest draining tank that has no inflow steam at all, Figure 1.1 (recognize the vagueness of the words *the same terms*—this is qualitative, and we will do better later as we work toward quantitative statements). Next, recognize that each of the two inflow streams might be described in terms similar to those representing the single inflow stream in Figure 1.2 (somehow, we would have to account for the different compositions of these streams and, in our case, the presence of only liquid in one of them and only solid in the other). Recognize further that we could expect to be able to find terms to describe (a) the mixing of salt and water, (b) the dissolving of the salt in the liquid, and (c) the flow of solid as well as the flow of liquid out of the tank.

Can you think of some even more complex examples that include all the components shown in Figure 1.3?

To go to the next step in the analysis of what is occurring in Figure 1.3, we would need mathematical statements to represent—*separately*—each of the phenomena or processes that we have recognized. This is a complex challenge; it may already be evident that we should expect to need a number of equations to represent what is occurring in Figure 1.3. For now, we defer this challenge and work instead (in the following chapter) on the simple case depicted in Figure 1.1, for which we will need only one equation.

Our first major quantitative task, addressed in the next chapter, is to find the equation.

Another Example: A Burning Candle

Let us now consider a chemical reactor (a device in which a chemical reaction or reactions occur), a burning candle. This example is complicated because chemical change (chemical reaction) is occurring; because three phases of matter (gas, liquid, and solid) are present; because melting of wax occurs; and because flow and mixing of gases take place. We proceed to analyze the burning candle in a *qualitative* way to understand how it works; in other words, we seek to resolve the various phenomena—that is, to list them separately—and these are both physical and chemical. As we do this, we ask questions such as the following:

- Why was the candle designed as it was?
- How were the materials chosen?
- How was the shape chosen?
- What are the fundamental phenomena that we can recognize?
- What experiments can we conceive to teach us how to analyze its performance and to provide quantitative information about it?

Components

Let us start with what is present in a burning candle that is cylindrical. Wax is the principle material, and there is a wick, a fibrous material such as cotton, which is present essentially as a thin cylinder at the central axis of the candle. In the flame, we see incandescent burning gases and infer the presence of burning wax and the co-reactant oxygen as well as the products of the reaction, expected to be carbon dioxide and water (and perhaps others, such as carbon monoxide and smoke).

Phases and Phase Changes

Next, what phases are present? By phase, we refer to gas, liquid, or solid. It is clear that the wax is present as both solid and liquid. We see them both, with the liquid in a pool at the top of the solid wax and possibly flowing (dripping) down the side of the candle. We also understand that the wax is vaporized and present in the gas phase near the top of the wick where combustion is occurring. The wick is a solid, and because it is (slowly) consumed in the burning process, we might reason that some of it could be vaporized as well. The reactants and products in the flame are present as gases.

Recognizing these three phases, we ask what phase changes occur. When the wax softens and melts near the hot flame, it makes liquid, and it vaporizes. When molten (liquid) wax drips down the side of the candle away from the hot flame, it cools down and resolidifies. Just by thinking about these points, we figure out that if the melting point of the wax is too low, the design of the candle will not be successful—the candle will bend or just melt and not work for long. And if the melting point of the wax is too high, the candle will not work either, as the wax will remain solid and not reach the wick effectively.

Flow

Looking for a connection to the draining tank, we ask what flows in the burning candle; we have already recognized the flow of liquid wax. We can also observe that gases are flowing. Just blow gently on the flame to increase the flow of air to it; the result will be evident in the flickering of the flame. Just put your hand above the flame, and feel the hot, flowing combustion gases as they rise.

Diffusion and Convection

Experienced engineers recognize that material moves not only by bulk flow processes such as those mentioned above (and in the draining tank), but also by molecular motion; this occurs in the gas phase, for example, in the candle flame, and it can also occur in liquid and solid phases. The process of transport by molecular motion is called diffusion, and it is not easily visualized in the candle.

To begin to understand diffusion, consider what happens when a drop of perfume is placed on a solid surface open to the air. The perfume evaporates, as molecules leave the surface of the liquid drop and enter into the air. There, the gas-phase perfume molecules move randomly as they collide randomly with air molecules and other perfume molecules. Because of the randomness of this process, the perfume molecules gradually spread out (diffuse) away from the drop and into the surrounding air. They are transported from a region of high concentration to regions of low concentration, just because the movements of the molecules are random.

We could also consider the movement of air molecules in the absence of perfume or any other added component. Although the air molecules are in random motion, there is no net transport of them because the process is random—the movements of molecules in one direction are balanced by the movements of the same kinds of molecules in the opposite direction; the result is that the molecules move around and bounce off each other chaotically, but the motions all average out so that there is no change in concentrations. We might imagine that all the N_2 molecules in a certain volume of air would move to the left while simultaneously all the O_2 molecules moved to the right, thereby leading to a change in local concentrations in the volume. If the number of molecules in the volume was extremely small, say, only a few, then this would happen sometimes, but if the number of molecules in the volume was large (say, the number in 1 mole, Avogadro's number, 6.023×10^{23}, which we write as a mol), then the probability of this occurrence is much too small to measure in an experiment.

The perfume molecules undergo a change in concentration because they begin in one place (in the drop) and diffuse into the surroundings. If they diffuse into a closed volume such as a room, then, eventually, sometime after the drop has evaporated completely, they become uniformly distributed throughout that volume, and the continuing random movements do not lead to any further change in concentrations at any position in the volume.

Because we can detect perfume molecules by smell, we can gauge approximately how fast their diffusion occurs, but we cannot quantify it very well, because we cannot quantify the amount of perfume very well with our noses.

Molecules in the flame of the candle similarly are transported by diffusion (as well as bulk flow), and they move quite fast because the temperature of the flame is high. The higher the temperature, the faster the molecules move.

Diffusion in liquids is much slower than diffusion in gases, because liquids are denser than gases and the collisions are much more frequent, so that the progress of the molecules is hindered. We cannot literally visualize the motion of molecules in a liquid, but we can observe the effect of this diffusion: think of what happens when a crystal of purple potassium permanganate is placed in a beaker of water. Because it is intensely colored, we see the result of the diffusion with our own eyes as the color spreads.

Diffusion in solids is even slower than diffusion in liquids, because solids are denser and more rigid than liquids. Some solids consist of atoms—think of a pure silver coin. Some solids consist of ions—think of NaCl crystals. Some solids consist of molecules—think of wax. Consequently, the species that move in solids as diffusion occurs depend on the nature of the solid. Diffusion of silver in the coin may occur primarily as silver atoms move into vacant positions in the crystal lattice of the silver; one can think of the diffusion as the motion of the vacancies.

To begin to quantify this comparison, E. L. Cussler, in his book *Diffusion: Mass Transfer in Fluid Systems* (second edition, Cambridge University Press, **1997**), wrote that diffusion in gases progresses at a rate of about 10 cm/min, whereas diffusion in a liquid progresses at a rate of about 0.05 cm/min and diffusion in solids at about 0.00001 cm/min.

Of course, molecules can move as a result of bulk flow, not just molecular motion. We could increase the rate of transport of the perfume molecules in the room by placing a fan next to the drop of liquid containing the perfume; the fan would force the air to move in bulk, and the fan would distribute the perfume much faster than it would diffuse. We would then say that the perfume transport is by both diffusion and *convection*.

Heat Transfer

How is heat (energy) transferred in the candle? We feel the hot gases flowing; they carry energy with them—by convection. Bulk flow carries both energy and matter. Diffusion occurs at the same time.

We also see that heat is transferred (or transported) not just upward with the convected (flowing) hot gases, but heat is also transferred from the hot flame to the wax; as the wax is heated by *conduction* of heat through it, some of the wax gets hot enough to melt. Conduction is a separate mechanism of heat transfer, different from convection. We could experience conduction by placing a metal spoon in a pot of boiling water and feeling how the temperature of the part of the spoon extending above the water surface increases as heat is conducted along the spoon. The end of the metal spoon would heat up faster than the end of a wooden spoon of the same size and shape, because metal is a good conductor of heat and wood is not. We call wood an insulator because it does not conduct heat well at all.

There is still another mechanism of heat transport illustrated by the candle—radiation. If we place a hand at the side of the flame, we can feel the transfer of heat by radiation, just as we feel radiation from the sun.

We have just considered several processes that we describe as physical processes: fluid flow, diffusion, and heat transfer. We call them *physical* processes as opposed to *chemical* processes. Chemical processes involve reactions (breaking and/or forming of chemical bonds); physical processes do not involve breaking and/or forming of chemical bonds—but chemical and physical processes occur together, as in the burning candle. Many processes important to chemical engineers involve both physical and chemical changes.

Chemical Reactions

What can we say about the chemical processes in the burning candle? The wax is consumed in reactions with oxygen from the surrounding air. If we approximate the wax as the alkane (paraffin) $C_{20}H_{42}$ (actually, wax is a mixture of alkane molecules of various molecular weights), then, knowing that the products of the combustion are primarily carbon dioxide and water, we can write the following stoichiometric equation to approximate the candle-burning reaction:

$$C_{20}H_{42} + 30.5\,O_2 \rightarrow 20\,CO_2 + 21\,H_2O \tag{1.1}$$

(Can you check whether the equation is balanced correctly?)

If our candle is a cylinder, can you figure out how we could measure the rate of the candle burning, using a stopwatch and a ruler? Why would we need the density of the wax to be able to represent the rate of this reaction in terms of molecules (or mols) per unit time? Why would we need to know the diameter of the candle? If we measured the rate of burning of wax molecules, how could we figure out the rate of formation of CO_2 molecules?

Steps toward Analysis of the Burning Candle

Here are some questions to consider about the candle, how it works, and why it is designed in the way it is:

- Can you think of an experiment that would allow us to check the stoichiometry of the wax burning reaction (to determine it quantitatively)?
- How might we make an approximate measurement of how much heat is given off in this chemical reaction?
- Why do you think the candle is approximately cylindrical?
- Why do you think that a pool of molten wax might remain at the top of the candle and not drip down the side?
- Can you suggest how the candle design might be improved by addition of a layer of wax with a high melting point surrounding an interior cylindrical section consisting of wax having a lower melting point?

WHAT WE MEAN BY QUALITATIVE AND QUANTITATIVE

We used the terms "qualitative" and "quantitative." Now let us explain them. Much of what engineers do is quantitative, that is, expressed in terms of numbers and/or equations. Engineers form the habit of making quantitative statements. To an engineer, a quantitative statement is often needed because a qualitative statement is often too vague to be useful. Suppose an engineer is assigned to analyze the performance of a bridge, for example, its stability in an earthquake. The engineer would naturally ask how long the bridge is. We might answer qualitatively that the bridge is long. This answer might seem meaningful and sufficient to some people, but not to an engineer who had to figure out answers to questions such as how much time would it take to walk across the bridge or what might it cost to retrofit the bridge for earthquake protection. The engineer who heard that the bridge is long would ask: long with respect to what? We could help and give a better engineer's answer by saying that the bridge is half as long as the Golden Gate Bridge. This is a specific, helpful, quantitative answer for someone who knows the Golden Gate Bridge, or who can quickly look up its length. We say that this answer is normalized—stated relative to some known standard. A more direct answer, the length of the bridge in meters, is also normalized because a meter is a standard length.

This normalized answer would be immediately helpful to an engineer who knew the Golden Gate Bridge and needed only an approximate estimate of the length of the other bridge; that engineer would have an immediate, easily visualized approximate understanding of how long the bridge is.

Can you think of some examples of representations of plants or animals that are represented in photographs including objects that provide some normalization? For a start, consider Figures 1.4 and 1.5. Can you estimate the length of the gecko or the volume of one of its eyes? What object in the figures provides the normalization? Read the figure captions to understand what the word "calibration" means. Which estimate (of the gecko length or eye volume) do you think will be more exact?

FIGURE 1.4   Adult male *Pachydactylus scherzi*, one of the smallest known geckos, found in the southern African country Namibia.

FIGURE 1.5   Calibration information for estimating the size of the gecko shown in Figure 1.4.

Engineers live by estimates and approximations, and they often work with approximations that are not very exact and proceed to make them more exact, as needed, but no more exact than is needed. Sometimes only rough approximations are sufficient, depending on the issues being addressed. An important part of engineering judgment has to do with how exact (or how approximate) a statement needs to be.

The most direct and informative answer to the question about the length of the bridge is quantitative: the bridge is 1,360 m long. Engineers typically make quantitative statements, and one of our goals is to help students develop this habit. But engineers also need to develop judgments about when such exact statements are needed and when rough (or even qualitative) statements are sufficient or better. Many people hearing quantitative statements regard them as dry and even inappropriate or difficult to comprehend. Many people would learn more from the answer that a bridge is half as long as the Golden Gate Bridge than from the answer that a bridge is 1,360 m long—because many people do not easily visualize how long 1,360 m is but do have a picture of the Golden Gate Bridge in their mind's eye.

ORDER-OF-MAGNITUDE STATEMENTS

Engineers and scientists often make statements that are quantitative but only approximate. Examples are order-of-magnitude statements, such as the following: the wavelength of the radiation is of the order of 10^{-8}m, or, alternatively, the wavelength of a different kind of radiation is of the order of 10^{-6}m. We say that these two values differ by two orders of magnitude, that is, by two powers of 10. Order-of-magnitude statements vary by powers of 10; so, for example, we would state that a wavelength of 3×10^{-6}m is of the order of 10^{-6}m, and a wavelength of 8×10^{-6}m is of the order of 10^{-5}m.

Rephrasing Cussler's statements about rates of diffusion, we would say that diffusion of a molecule in a gas is two orders of magnitude faster than that in a liquid and six orders of magnitude faster than that in a solid.

Order-of-magnitude estimates are sometimes sufficient for engineering purposes. Order-of-magnitude statements are sometimes called semiquantitative.

WHAT DO WE MEAN BY SMALL OR LARGE?

Sometimes quantitative statements lend themselves to simplification by neglect of something by comparison with something else. Consider the use of a steel ruler to measure the length of a plank of pine wood. We may make the measurement in the winter when the temperature is 0°C or in the summer when it is 30°C. The lengths will not be exactly the same, because the ruler and the wood expand as the temperature increases, and they do not expand to the same degree, because the metal and the wood have different coefficients of thermal expansion. We might say that, for a construction project, we need to know the length of the plank of wood within an uncertainty (or tolerance or error) of 1%. We can look up the coefficient of thermal expansion of steel and that of pine, finding them to be approximately 12×10^{-6} and 4×10^{-6} m/(m × °C), respectively (the latter along the grain of the wood). The difference is about 8×10^{-6} m/(m × °C), and for a 30°C change in temperature about 2.4×10^{-4} m/m, or about 0.024%. If our goal is to determine the length of the plank with an uncertainty of 1%, then we would say that the error (uncertainty) associated with the temperature change is 0.024/1, or only 2.4% of the tolerable uncertainty. As engineers, we might conclude that this uncertainty is small enough to neglect when our tolerance (which is two orders of magnitude greater than this) is considered. An important point is that we do not simply say that the error associated with the temperature difference in the measurements is small—we say that it is small in comparison with our tolerable uncertainty. In other words, we quantify the statement by normalizing it. Whenever engineers say that something is small or large, they need to normalize it by saying small or large by comparison with something else—and the something else is determined by engineering judgment.

An engineer would ask whether the uncertainty associated with the temperature-induced changes in ruler and plank lengths is small or large by comparison with the uncertainty associated with our measurement of the length using the ruler: how exactly can the measurement be made with the ruler and our eyes? Is the uncertainty in the measurement associated with the expansion of the steel and wood small by comparison with the uncertainty associated with the measurement using the ruler and our eyes?

When we are justified in saying that some observation or estimated result is small enough to neglect (is negligible) by comparison with another—with the comparison quantified and the meaning of "negligible" explained in our context—we proceed without taking further account of what we have shown to be negligible. When the error associated with the temperature dependence of the steel and wood is negligible with respect to the error in length measurements with the ruler and eye, we do not consider it further. Engineers often develop the experience to determine what is negligible without having to check carefully.

STEADY-STATE AND TRANSIENT PROCESSES

The tank-draining and candle-burning processes were ones in which characteristics of the system changed with respect to time. The tank drained; the volume of water in it changed with time. The candle burned; the height of the candle (the amount of wax remaining) changed with time. Processes like these in which a property or properties change with time are called *transient* or *non-steady-state*. These points are illustrated in Figure 1.6 for the draining tank and for a filling tank.

Now consider a tank that is being filled as it drains (Figure 1.7). It is possible to choose a rate of flow of water into the tank to just match the rate of draining. In this case, the liquid level in the tank does not change with time. This is not a transient process. Instead, it is called a *steady-state* (or stationary-state) process (with respect to the tank). If the flow rate of water into the tank or the flow rate of water out of the tank would be changed from the steady-state values, then the liquid level in the tank would change with respect to time, and the process would become transient.

Engineers work with many processes that operate normally at steady state or nearly so. An automobile being driven at a constant speed on a straight, level road operates at approximately a steady state. But the automobile must go through

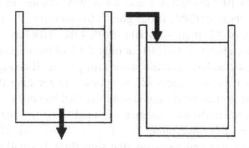

FIGURE 1.6 Example of transient processes, a draining tank (left) and a filling tank (right). In each of these, the liquid level changes with respect to time.

FIGURE 1.7 Tank with inflow and outflow streams of water. This could be operated at a steady state, and then the flow rate of water into the tank would equal the flow rate of water out of the tank. When these flow rates are not equal, the liquid level in the tank changes with respect to time, and the process is transient (or non-steady-state).

a transient operation to get to this constant speed and near-steady-state condition. After being started, it would undergo changes in speed and accompanying changes in the conditions in the engine, for example, being heated up. After some time, the operation could approach the steady state. And after some time, the fuel would be fully consumed, and the automobile would slow down and come to a stop during a period of transient operation. When stopped, it would be in another steady state.

Often processes are designed to operate stably at steady state, and steady state is considered the normal operating mode. But any process has to undergo a transient period of operation before it gets to steady state. Start-ups and shut-downs are transient operations, and so are unforeseen upsets such as those caused by power outages, leaks, and plugs that may form and hinder flows.

EQUILIBRIUM AND CONTRAST TO STEADY STATE

Consider two tanks on a level plane connected by a pipe connected to the base of each, with a valve in the pipe initially closed. If one tank is initially full and the other initially empty, then when the valve is opened, the fluid will flow from the filled to the empty tank until the levels in the two tanks are the same. This final state corresponds to equilibrium. It is reached through a transient process and reaches a steady state, which is the equilibrium state. In contrast, the example of Figure 1.7 corresponds to a steady state but not an equilibrium.

A dictionary definition of equilibrium is a state of balance between opposing forces or actions—the force for driving fluid from the initially filled tank was initially higher than that from the empty tank (none), but, when equilibrium is reached, the equally filled tanks have the same driving force for flow of the fluid, and so there is no flow.

In chemistry, equilibrium in a multiphase mixture such as liquid water and ice means that the rate at which liquid water freezes is matched by the rate at which ice melts. Equilibrium in a chemical reaction means that the rate at which the reactants are converted into the products is equal to the rate at which the products are converted into the reactants. The forward and reverse reactions are in balance, characterized by equal driving forces.

WHAT IS CHEMICAL ENGINEERING?

The concepts and methods introduced in this book relate to the profession and field of chemical engineering. Chemical engineers discover, invent, analyze, design, innovate, and facilitate the creation of processes and the manufacture of products by processes that involve chemical and physical change—that is, those involving chemical reactions and/or those that take place in association with them, such as fluid flow, mixing of fluids and solids, heat transfer, and separation of components in mixtures. Historically, chemical engineering arose from the practice of chemical manufacture; we might trace it back to wine making in ancient Egypt. Early chemical engineers worked with discoveries in chemistry and other fields, and figured out how to apply them on a large scale. Today chemical engineers continue to work on technologies for manufacture of chemicals and fuels, but they have expanded their field to include protection of the environment, biological processing, and medicine. Some of the examples and problems in this book introduce some of these applications.

RECAP AND REVIEW QUESTIONS

We have now begun to understand how to consider what happens in situations involving physical and chemical changes. We categorize the changes in terms of flow, mixing, dissolution, heat transport, diffusion, phase change, and chemical reaction, and often ask what phases are present, what phase changes take place, and what reactions take place. In the simplest cases, only one of these types of processes occurs, as illustrated by the draining tank (which we describe in terms of a single flowing stream). But in complex cases, illustrated by the burning candle, multiple processes take place simultaneously. In analysis, we strive to separate them and consider them one at a time. This idea of going from a very simple to increasingly more complex situations is illustrated by going from a tank with one outflow stream; to a tank with an inflow and an outflow stream; to a tank with two inflow streams and one outflow stream; and to a tank in which salt is dissolving as the tank contents are mixed and there are inflow and outflow streams.

What we have done in this chapter is almost entirely qualitative, that is, done with words alone and without numbers or equations. The equations appear in the next chapter, for the simplest of our examples, the draining tank. Then we proceed in later chapters to address more complex situations quantitatively.

Test your understanding of the terms used here by describing qualitatively the transient operation of a tank with inflow and outflow streams and the near-steady-state operation of a candle. Contrast steady state to equilibrium and give some examples to illustrate their meanings. Write qualitative, quantitative, and normalized statements to describe a burning candle. Give examples of physical processes and how to do experiments to visualize them. Do the same for chemical reactions. Contrast physical processes and chemical processes.

As you read the chapters following this one, come back and review this chapter for ideas that relate to how fast some particular change is taking place.

PROBLEMS

1.1. Prepare data sheets to be used in the laboratory to measure the rate of draining of water from tanks. Each tank is mounted with an orifice at its base. See the following schematic representation:

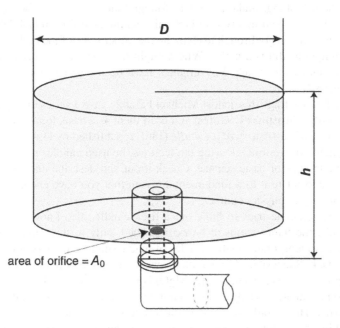

The data sheets should facilitate rapid recording of heights of liquid above the orifice, h, and the time after the flow of water starts, t (as well as the appropriate dimensions of the tank, which are shown on the schematic drawing). Note that the scale mounted on the side of the tank in the laboratory is not h. Good laboratory data include such information as who did the experiments, when and where the experiments were conducted, and notes about how the measurements were made—so that other people can reproduce them. Be sure to include units of the measurements.

1.2. Go to the library or online and find the *Kirk-Othmer Encyclopedia of Chemical Technology* (Kroschwitz, J. I.; Seidel, A..; Kirk, R. E.; Othmer, D. F; *Kirk-Othmer Encyclopedia of Chemical Technology*; John Wiley; 2004.) and also *Ullmann's Encyclopedia of Industrial Chemistry* (Ullmann, F. and Bohnet, M.; *Ullmann's Encyclopedia of Industrial Chemistry*; John Wiley; 2009.). Look up the entries on wax, do some further digging, and find out what paraffin (or paraffinic) wax is and how it is manufactured from coal by way of coal gasification and Fischer-Tropsch synthesis. What are the chemical reactions in coal gasification and Fischer-Tropsch synthesis?

  A. Write a summary of these facts, no more than one-half page.

  B. Determine an estimate of the melting temperature of paraffin wax suitable for use in candles.

1.3. Consider a sieve consisting of a cylindrical container with a flat base in which there are *n* holes of equal diameter. Describe how you would start to analyze the process of water draining from the sieve. How would you predict the rate of water flow relative to that in a tank with the same dimensions but only one hole (of the same size as those in the sieve) in the base?

1.4. Without looking, make an order-of-magnitude estimate of the volume of liquid that is held in a typical U.S. railroad tank car for ethanol. What reasoning did you go through to estimate the dimensions of the tank car? What assumptions did you make? Where might the ethanol have been manufactured and from what? Why might it have been destined for a petroleum refinery?

1.5. The famous English scientist Michael Faraday gave a series of popular lectures (the Christmas Lectures) and used them as a basis for a book entitled *The Chemical History of a Candle* (1861; republished by Dover, 2002; also available free online). During the lectures, he used candles to demonstrate some lessons of basic science. Check it out and describe one such experiment and relate it to a fundamental lesson that you have encountered in a chemistry or physics course.

1.6. When someone tries to blow out a trick candle, also known as a magic candle, the flame seems to be extinguished, only to flare up again. Trick candles incorporate small particles of magnesium in the wax embedded in the wick. Suggest an explanation for what the magnesium does.

1.7. High-purity magnesium oxide (MgO) is manufactured by burning of magnesium. Suggest briefly a conceptual design of a process for making MgO powder. How would you make your process a continuous process, that is, one in which MgO powder is produced continuously at a steady state?

1.8. In an article in the journal *Chemical and Engineering News* (August 9, 2010, p. 34), it is stated that candle wicks are usually braided cotton treated with a chemical salt. Find this article to answer the question, what is the salt and what is its purpose in the design?

1.9. Check out why whale oil was so important to people before petroleum products became widely available, and write a short summary of your findings.

1.10. Design an oil-burning lamp; sketch the design. Explain briefly how it works and what it has in common with a candle. How is it different from a candle? What can you learn about oil-burning lamps used in antiquity? Can you find one to purchase online? Going well beyond the subject of this chapter, can you find out how the age of such a lamp can be estimated?

1.11. Do some research and prepare a short summary of each of the following, pointing out the contributions of chemical engineers and those who preceded them:

A. Wine making and its history.

B. Ammonia manufacture from nitrogen and hydrogen. See, for example, the book *The Alchemy of Air*, by Thomas Hager (Three Rivers Press, 2008).

C. Ziegler–Natta catalysis and the manufacture of plastics.

D. Penicillin and its manufacture.

E. The artificial kidney.

  F. Catalytic converters in motor vehicles and their impact on air quality.

  G. Chemical warfare agents.

  H. Gold mining with cyanide.

  I. Directed evolution.

1.12. Do some research and find out why the name Bhopal is so significant to chemical engineers.

1.13. Do some research and make some order-of-magnitude estimates of the ratio of the volume of a mol of methanol in the gas phase (at atmospheric pressure) to that in the liquid phase to that in the solid phase. Choose temperatures for the comparisons and explain why you chose them.

1.14. Do some research and make some order-of-magnitude estimates of the ratio of the viscosities of the following liquids at room temperature: water, methanol, honey, and commonly used automobile engine oil ("motor oil"). Make the same comparisons when the temperature is 1.0°C.

1.15. Do some research and make some order-of-magnitude estimates of the ratios of the volumes of the planets surrounding the sun. Do the same for the ratios of masses of the planets. Add the earth's moon; the eight largest moons of Saturn; and the four largest moons of Jupiter to the results.

1.16. Do some research and make some order-of-magnitude estimates of the ratios of the volumes of the following adult animals: a carpenter ant, the gecko *Pachydactylus scherzi* (depicted in this chapter), the largest tortoise found in the Galápagos Islands, the largest African elephant, and the largest known dinosaur.

1.17. Do some research and make some order-of-magnitude estimates of the ratio of the surface area to the volume of the following adult animals: a carpenter ant, the gecko *Pachydactylus scherzi* (depicted in this chapter), the largest tortoise found in the Galápagos Islands (with and without the head and legs withdrawn into the shell (carapace)), the largest African elephant, and the largest known dinosaur.

1.18. Do some research and find out what perfluorooctanesulfonic acid is. What is it used for? Why is the presence of this compound in the environment important? How does it get there? How is it most likely to be consumed by people? Why is it harmful to people? What is the U.S. Environmental Protection Agency, and why might people in that agency be paying attention to research on perfluorooctanesulfonic acid in the environment?

1.19. Find out what the term "planetary boundaries" means in the context of chemical and fuel processing, and find some examples of how concern for planetary boundaries has come into play in planning for the design of processes involving earth-abundant raw materials such as natural gas, petroleum, coal, and biomass.

2 Analysis of a Draining Tank

Experiments and Equations

ROADMAP

This chapter illustrates the first application of quantitative reasoning for analysis of a simple physical process, draining of a tank through a hole in its base. We review graphical forms of some familiar equations and look ahead to comparing plots of data for a draining tank as a start to finding equations that are good candidates to represent the data. In this analysis, we are guided by physical reasoning, and this chapter illustrates how mathematical and physical reasoning are combined. The concepts of differential calculus are central to this chapter—it is concerned at its core with the *rate* of the tank draining. The analysis presented here is used as a foundation for the design of tanks and systems of tanks, including a tank designed to give a constant flow rate for extended periods.

ENGINEERING ANALYSIS

The draining tank mentioned in Chapter 1 provides an ideal starting point for illustrating the methods of engineering analysis. Our goals are to represent—*quantitatively*—the draining process, the flow of water through the orifice at the base of the tank. Our goal is to find an equation that represents tank-draining data—with the appropriate terms in the equation (the parameters) determined quantitatively.

The resulting equation is known as a mathematical model (or, more commonly, just a model). The word *model* implies that we will use a simplified representation of the physical phenomenon, an approximation. An important part of engineering judgment involves choices of models that are both simple enough to be used effectively and accurate enough (not too simple) to represent reality with sufficient reliability. A good model must represent the essential characteristics of the physical reality. What is essential and what is sufficiently reliable depend on the application and require engineering judgment.

It is important to realize the limitations of any model. Every model is a simplification, an incomplete representation of reality. The statistician G. E. P. Box is known for the quotation, "All models are wrong, but some are useful." As engineers use models, they need to realize their limitations.

In all engineering activities, we strive for the most appropriate degree of approximation—this is sometimes called the principle of optimum sloppiness. We economize our resources, including our time. We do the job just well enough, or somewhat better, building in a cushion to guarantee success. Doing it much better than well

DOI: 10.1201/9781003429944-2

enough would be an inefficient use of resources; doing it less than well enough may mean failure or even a lack of responsibility, such as in matters of safety and environmental protection.

Often an engineer starts a project with a quite rough set of approximations, arriving at some preliminary conclusions. Those conclusions may make clear that there is no point in going further with the project—it may just be too unsafe or too costly to make sense. On the other hand, if the preliminary conclusions are encouraging, then the engineer may put in more effort, using less-rough approximations and making progress toward a more reliable set of conclusions. If these are still encouraging, then the engineer may continue to improve the approximations step by step, until finally they are good enough for a reliable set of predictions for the final product or process. The basis for improving the approximations often involves experimentation combined with calculations, with increasing attention to detail and consultation with colleagues who can provide needed experience and expertise (and ask critical questions).

When we make approximations, we need to know what they are and how they limit our analysis. In any analysis, an engineer must decide what issues are too insignificant to matter and which are significant and essential. This point was illustrated in Chapter 1 with the example of the ruler and the measurement of the length of a wood plank.

The example that is central to this chapter is the draining of water from a tank. The draining is caused by the force of gravity acting on the water—driving the flow. The force of gravity varies from one position on the earth's surface to another, and so the draining of a tank in one place may be different from that in another. But the variations in the force of gravity between locations are usually so small *relative* to other sources of uncertainty in the measurements of flow that the variations in the force of gravity are negligible relative to these other sources of uncertainty. We stress the word *relative*; this statement is normalized; the variations in the force of gravity are small in comparison with the uncertainties (errors) in the measurements. Errors are intrinsic to any measurement, and we return to them as we develop our subject.

Now let us be more specific about the meaning of *engineering analysis*. Its definition is related to the definitions given in Chapter 1, but it is specialized to engineering. Engineering analysis involves resolution of observable phenomena into fundamental components and representation of these phenomena in mathematical terms, as illustrated throughout this book. Such analysis also connotes a comparison of the observations with the mathematical representation (i.e., comparison of the data with the model) and manipulation of the mathematical expression to predict the behavior in related situations. A statement of a model is improved by a statement of its limitations. This definition of analysis goes well beyond the dictionary definitions of analysis stated in Chapter 1.

Engineering Design

The predictive element built into this definition of engineering analysis is part of the essence of engineering—engineers use analysis as a basis for prediction of behavior of observable systems, and also for creation of new devices and systems and processes—that is, for *design*. Design is a central part of engineering, and the challenges and rewards of observing successful designs being used are what draw many engineers to their profession.

How is the verb *to design* defined? A dictionary definition is to plan; to make sketches or drawings for. An architect's definition would be similar and would include drawings with quantitative information (dimensions) and specification of materials and notes about suppliers and methods of construction. The engineer's definition of design is often even more inclusive than this. In engineering, to design is to prepare the basis for creating or producing devices and products and processes that often have many intricate interconnected components. The basis must be quantitative, and often it is highly detailed. Design is possible only on the basis of analysis. Designs are often the work of teams.

GETTING STARTED WITH ANALYSIS

How do we begin to analyze a physical situation? There are various approaches:

1. Do experiments (make observations) and then find equations (models) to represent the observations (data) by finding out what fits well.
2. Use theory based on scientific principles as a starting point (but remember that all theories are ultimately based on experiment and are limited too).
3. Use some combination of these two approaches and reject any model that violates fundamental principles.
4. Use guidance from dimensional analysis.

This list is only a start, but it is a good jumping off point into our example of the draining tank. We begin with the first approach (and it is helpful if students have gone into a laboratory and collected their own data) and then work down the list as we go beyond this chapter.

In the laboratory, we measure the height of liquid above the orifice mounted at the base of the cylindrical draining tank (call this height h) as a function of the time after the draining starts (call this time t). We seek to find an equation, a model, to express h as a function of t, and then we plan to manipulate the equation to determine the flow rate of water out of the tank as a function of h and of t. Such information can be generalized to provide the basis for design of draining tanks other than the one used in the laboratory.

The approach that we are now taking is called empirical, meaning that it is based on observation and not theory. We seek to find an equation that fits the data well, with our basis for choosing the equation being the goodness of fit. The word *fit* is what indicates the empirical nature of the approach. But, even at the outset, we go beyond simple empiricism and use physical reasoning to recognize some necessary characteristics of the equation.

To start with this physical reasoning, we realize (a) that it is the force of gravity on the water that causes the flow and (b) that the higher the pressure on the water flowing out of the tank at the orifice, the higher the flow rate of the water. This pressure (called the head) depends on h, and we therefore recognize that (*i*) our equation will show a decreasing flow rate with deceasing values of h (i.e., as the tank drains) and (*ii*) as a limiting case, as h approaches zero, the rate of flow will also approach zero—in other words, the driving force for flow approaches zero as the height of liquid above the orifice approaches zero—as this limit is approached, the amount of water left to drain approaches zero.

To repeat, we recognize at the start that the flow rate of water will be a maximum when the tank is full and will decrease as the tank drains, approaching zero as the liquid level approaches zero. Our equation will have to account for these statements, and so we have guidance in the choice of the equation. We will reject any candidate equations that do not meet these physically evident criteria.

Notice how we are reasoning in terms of the *rate* of flow—visualizing the draining tank, we have a physical conception of what the rate means: how fast the water flows out of the tank. Our approach to the draining tank analysis starts off with rates, and not just with observations showing how h depends on t, which provide less insight. We will see that much of engineering analysis starts from consideration of rates, illustrated in the examples stated in Chapter 1 mentioning rates of heat transfer, rates of diffusion, and rates of chemical reaction. These statements about rates should alert students to a central point: we will use the fundamental concepts of calculus.

To fit our data to an equation, we need to find an equation that is a good approximation of the data—this statement explains what we mean by a good fit. To proceed, we will examine the data and postulate candidate equations to represent them. Thus, we need some skill at *reading* (or *visualizing*) *data* and *reading equations*. When we examine data, we should think about what form of equation would represent them well. To develop the skill of reading equations, we need to see connections between equations and the shapes of the lines on graphs that represent them.

READING EQUATIONS AND GRAPHS

To start, let us read some simple equations and the corresponding plots. As we proceed, we assume that the function represented on the vertical axis (the ordinate) of the plot is y, and that y is a function of x, which is plotted on the horizontal axis (the abscissa); thus, $y = f(x)$. We begin by representing the plots on graphs with linear scales (later, we will use nonlinear scales (logarithmic scales)). As we read the equations and graphs, we make it a habit to look for *limiting cases*, as they help us to understand and visualize the equations and graphs.

Consider the following examples, starting with the simplest:

Equation	**Plot**

1. $y = $ constant
(for example, $y = 1$)

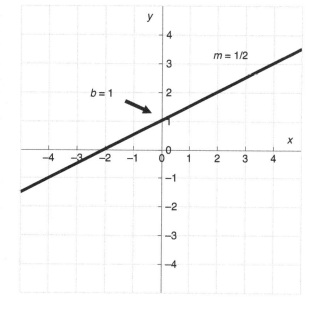

2. $y = mx + b$
(another straight-line plot)

3. $y = x^2$
(a parabola)

4. $y = x^3$

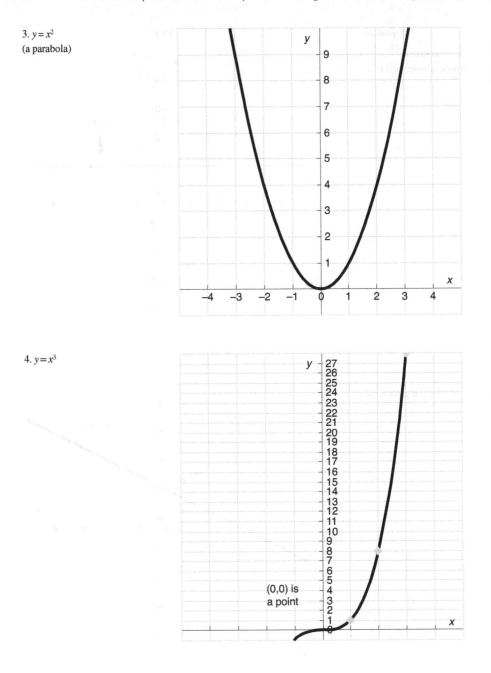

(0,0) is
a point

5. $y = x^{1/2}$
(shown only for $x > 0$)

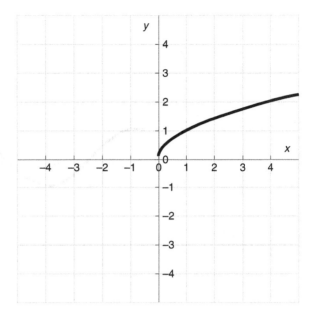

6. $y = \cos(x)$

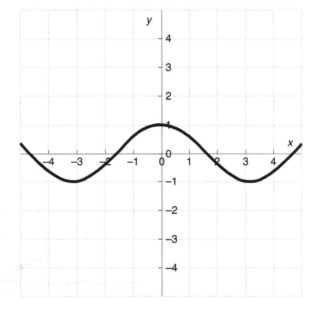

7. $y = \sin(x)$

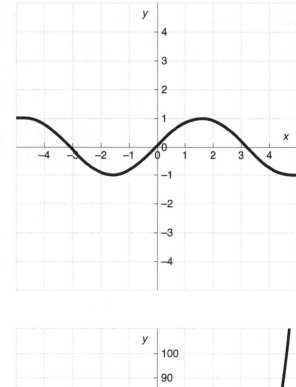

8. $y = e^x$

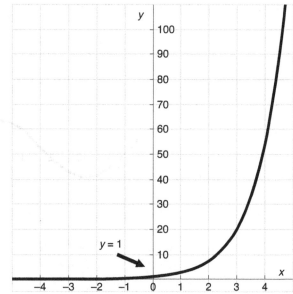

9. $y = \log_{10}x$
(or $10^y = x$)

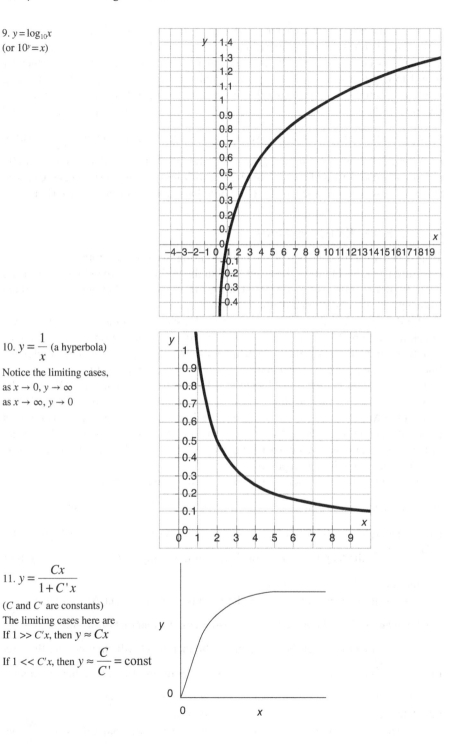

10. $y = \dfrac{1}{x}$ (a hyperbola)

Notice the limiting cases,
as $x \to 0$, $y \to \infty$
as $x \to \infty$, $y \to 0$

11. $y = \dfrac{Cx}{1 + C'x}$

(C and C' are constants)
The limiting cases here are
If $1 \gg C'x$, then $y \approx Cx$

If $1 \ll C'x$, then $y \approx \dfrac{C}{C'} = \text{const}$

To repeat a point from Chapter 1, in the consideration of the limiting cases of this equation, it is not sufficiently meaningful to say that when x is small, $y = Cx$ or that at when x is large, $y =$ constant. To make such statements quantitative and thereby useful to engineers, we need to specify small or large with respect so something else. For example, in considering the imprecise statement that if $C'x \ll 1$, then $y \cong Cx$, we can be more precise by stating what we mean by the approximately equal sign ($\cong$, sometimes written as $\approx$). We might say that Cx must match $\dfrac{Cx}{1+C'x}$ within an error (deviation) of no more than 3%, for example. If the errors in experimentally measured values of x and y were about 3%, this might be a good statement. Can you calculate how much smaller than 1 $C'x$ would have to be to meet this criterion?

AVERAGE RATES, RATES, AND LIMITS

To proceed with the analysis of the tank draining, we consider rates of water flow (measured by rates of fall of the water level in the tank). Let us reemphasize what we mean by rates—now using concepts familiar from calculus, but approaching them from a physical rather than a mathematical viewpoint.

Our data provide values of h as a function of t, sketched in Figure 2.1a and shown in detail in Figure 2.1b. In Figure 2.1a, we have drawn a curved line to pass approximately through our data points (which are not shown to keep the drawing simple); the line was drawn by eye as a first approximation to represent the data.

Let us consider the *average* change in h for a particular change in t. We start by picking an arbitrary anchor point, shown in Figure 2.1a. To find the average change in h, called Δh, in a particular time interval, called Δt, we read off the two points on the right triangle intersecting the curve at the top of the vertical edge and the right of the horizontal edge; we construct the triangle shown by drawing the straight line (chord) connecting these two points.

In the time interval chosen, h decreased, so that the change in h, Δh, is negative, and t has increased, so that Δt is positive. Let us call $-\dfrac{\Delta h}{\Delta t}$, which is positive (or zero), the average rate of fall of h in the time increment Δt. On the graph of Figure 2.1a, $-\dfrac{\Delta h}{\Delta t}$ is $-\dfrac{\text{rise}}{\text{run}}$, the length of the vertical edge of the triangle divided by the length of the horizontal edge of the triangle.

Now, using the same anchor point, let us make the increment Δt smaller; we see that the magnitude of the increment in h also is smaller and that the average rate $-\dfrac{\Delta h}{\Delta t}$ for this increment is larger. Next, we make the increment still smaller, and the rise/run ratio $-\dfrac{\Delta h}{\Delta t}$ is still larger. Let us do this for a series of smaller and smaller increments in time. The result is a graph such as that sketched in the top part of Figure 2.2 (the data are shown in the lower part of this figure).

In calculating the points for Figure 2.2, we recognize that our values of $-\dfrac{\Delta h}{\Delta t}$ get harder and harder to determine accurately as Δt gets smaller and smaller (because the increments become so small that it becomes difficult to make the measurement

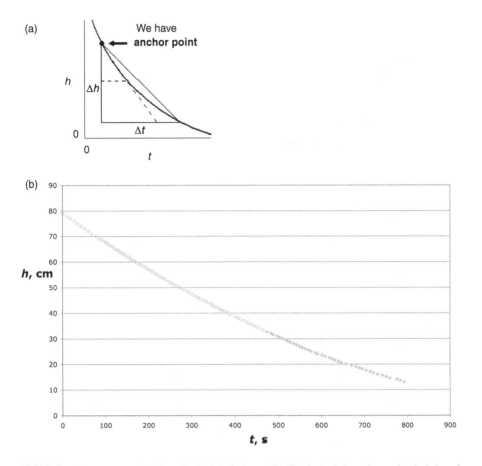

FIGURE 2.1 (a) Sketch of tank-draining data. (b) Tank-draining data: the height of water above the orifice as the tank drains. Each point (square) on this graph represents a measurement made in the laboratory. The open squares were recorded by one group of experimenters and the closed squares by another. The line sketched in (a) approximates the points on (b).

on our graph). Hence the uncertainty (error) in $-\dfrac{\Delta h}{\Delta t}$ becomes larger and larger as Δt approaches zero. We also see from the graph that the value of $-\dfrac{\Delta h}{\Delta t}$ converges toward a particular value in this limit as Δt approaches zero.

Thus, we recognize that we cannot literally measure $-\dfrac{\Delta h}{\Delta t}$ at the smallest values of Δt, because both Δh and Δt are too small to measure on the graph, but we also recognize that the value of $-\dfrac{\Delta h}{\Delta t}$ converges to a value that we can find by *extrapolation*, that is, by extending the curve fitting the $-\dfrac{\Delta h}{\Delta t}$ data vs. the Δt data to the limit it approaches as Δt approaches zero, as shown on Figure 2.2.

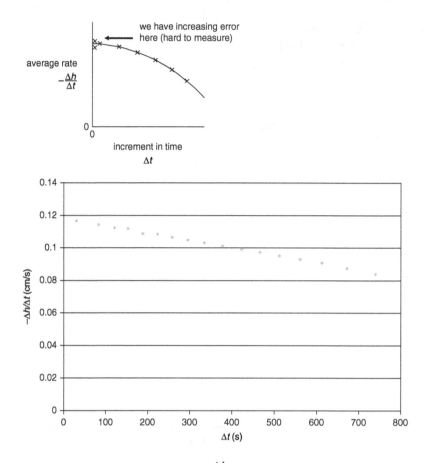

FIGURE 2.2   Average rate of fall of h, $-\dfrac{\Delta h}{\Delta t}$, increases toward a limiting value as Δt approaches zero. The upper graph is just a sketch; the lower one shows the data.

Returning to Figure 2.1, we recognize that the values of $-\dfrac{\Delta h}{\Delta t}$ determined for each value of Δt correspond to the negative of the slope of the chord connecting the two points on the curve representing the data corresponding to the particular values of Δh and Δt. As Δt gets smaller, the magnitude of the slope of the chord increases to a limiting value. That limiting value, as Δt approaches zero, is just the slope of the tangent line to the curve representing the h vs. t data at the anchor point.

The value to which $-\dfrac{\Delta h}{\Delta t}$ converges is called $-\dfrac{dh}{dt}$; to repeat, this is the slope of the curve showing h vs. t at a particular value of t (and the corresponding value of h).

Now, with $-\dfrac{dh}{dt}$, we are no longer considering an average rate. Instead, we have found a *point* value of the rate, the rate of fall of the liquid level at a particular time. By finding the slope of the tangent line of the curve at any point, we find the rate of fall of the liquid height at that point. So, for any measured height or time, we can determine this rate by drawing the tangent to the curve and measuring its slope.

In terms of calculus, we write the following:

$$\lim_{\Delta t \to 0} -\frac{\Delta h}{\Delta t} = -\frac{dh}{dt} \qquad (2.1)$$

and $-\dfrac{dh}{dt}$ represents the first derivative of h with respect to t (multiplied by -1). Mathematically, if h is a well-behaved (differentiable) function of t (single valued and with no discontinuities in the h vs. t curve that is presumed to exist)—then, for each value of t, there is one value of h, and we can determine $-\dfrac{dh}{dt}$.

Does it seem plausible that, once the flow has started and before it stops, h is a smooth, continuous, single-valued function of t? One reason why we chose the example of the draining tank is that observation of the draining tank makes this statement seem plausible—the flowing water as we visualize it seems to be flowing continuously in a steady stream.

If we had an equation for h as a function of t, we could use calculus to evaluate $-\dfrac{dh}{dt}$ at any value of t. We are on the way to having such an equation.

The rate $-\dfrac{dh}{dt}$ can be related to the flow rate of water out of the tank, as follows. The tank is assumed to be cylindrical, with a constant cross-sectional area A, and so the volume of water in the tank is A times the height of water above the base of the tank. We call this volume V. Now, because A is constant for a cylindrical tank,

$$-\frac{dV}{dt} = A\left(-\frac{dh}{dt}\right) = \text{volume flow rate} \qquad (2.2)$$

We have seen how to determine the volume flow rate of water from the tank at any measured value of h, but did we literally measure this rate? No, we determined it only indirectly, using our data for h vs. t to evaluate it, by taking slopes of the h vs. t curve and knowing how to relate h to V.

We say that we differentiated our data to evaluate this rate.

ANALYSIS OF TANK-DRAINING DATA: EVALUATION OF RATES

To proceed with the analysis of the tank-draining data, let us use our plot of h vs. t to determine values of $-\dfrac{dh}{dt}$ at various values of h. The data shown in Figure 2.3 were determined in this way.

Let us consider further why it is helpful to look at the data this way. We know from our observations that h depends on time, and we could find an equation to express how, $h = f(t)$. But we also realize that the cause of the flow is the force of gravity on the water in the tank. Our physical understanding—that the greater the height of

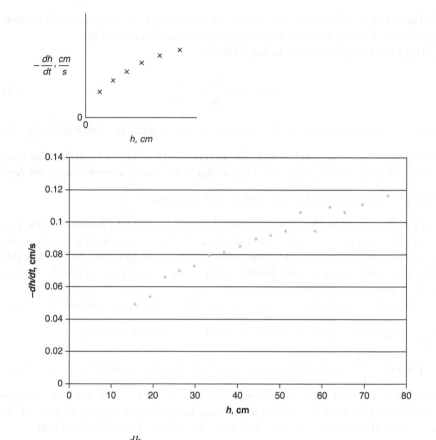

FIGURE 2.3 Values of $-\dfrac{dh}{dt}$ determined from the h vs. t data, shown as a function of h. The upper plot is just a sketch; the lower one includes the data.

liquid above the orifice (h), the greater the flow rate of the water—leads us to the realization that there is something fundamental and correct about seeking a relationship between the flow rate and h, as we are accounting for the cause of the flow.

The data in Figure 2.3, of course, confirm the expectation that the flow rate increases with h. Now, the next point is that our understanding of the physical situation helps us draw a curve to fit (represent) the data in the plot. Specifically, we know that as the height of liquid above the orifice approaches zero, the flow rate of liquid through the orifice approaches zero.

And so, to start, we use the data and draw the curve passing through the origin in the graph shown in Figure 2.4.

We might have been tempted to represent the data with the straight line shown in Figure 2.4, because it fits the data quite well. But we know that this is wrong—because it does not represent the limiting case that we recognized in our consideration of the physical situation, namely, that the intercept of the proper curve fitting the data must be at the origin of the graph. In other words, we have made a significant

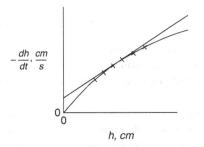

FIGURE 2.4 Sketch of the flow rate data of Figure 2.3 showing both a straight line and a curved line through the data. The latter is a good choice because it shows the proper limiting behavior by passing through the origin.

advance by using this result: $\lim_{h \to 0} -\dfrac{dh}{dt} = 0$. Therefore, we see that there must be substantial curvature in the plot, because it must go through the origin.

Now, we read the graph with the curved line representing the data in Figure 2.4 and ask: What equation is a good candidate to represent this line? Looking at the equations we considered above, we see that the best of them is $-\dfrac{dh}{dt} = C''h^{1/2}$ (where C'' is a constant). But we recognize that similar graphs would have been obtained for plots of the form $y = C'x^{1/3}$ or $y = C'x^{2/3}$, for example. How can we narrow down our estimate of the value of the exponent on x?

To make this step forward, we recognize the advantage in plotting our data in a way that will determine an estimate of this exponent directly. We thus take advantage of logarithmic functions, which have the following properties:

$$y = ab \tag{2.3}$$

$$\log(y) = \log(a) + \log(b) \tag{2.4}$$

or

$$\ln(y) = \ln(a) + \ln(b) \tag{2.5}$$

$$y = ab^n \tag{2.6}$$

$$\log(y) = \log(a) + n \log(b) \leftarrow \text{this is of the form } y = mx + b \tag{2.7}$$

or

$$\ln(y) = \ln(a) + n \ln(b) \leftarrow \text{this is of the form } y = mx + b \tag{2.8}$$

And so, for the tank draining, we postulate $-\dfrac{dh}{dt} = C''h^n$ (where C'' is a constant) and evaluate n. Thus, we make a plot of $-\dfrac{dh}{dt}$ on a logarithmic scale vs. h on a logarithmic scale (this is a log–log plot). (It is not proper to make a plot of the logarithm of $-\dfrac{dh}{dt}$ vs. the logarithm of h, because it is mathematically incorrect to take the logarithm of anything but a pure number.) (Do you recall what logarithmic coordinates are? Can you download graphs or find graph paper with logarithmic coordinates?) The desired plot is shown in Figure 2.5.

The slope of the straight line providing a good fit to the data in Figure 2.5 is nearly 0.50 (can you verify that?), and therefore we make the approximation that $n = \frac{1}{2}$. Instead of taking n to be exactly one half, we might have taken the exact value of the slope of the straight line fitting the data of Figure 2.5. There are several reasons why we do not, and we point them out below at a more logical stage of the analysis, where we go beyond empiricism guided by physical reasoning and invoke theory.

Now, to complete the equation and make it fully quantitative, we need to determine the value of the constant C''. One simple way to do this is take a single value of $h^{\frac{1}{2}}$ and the corresponding value of $-\dfrac{dh}{dt}$ and calculate C'' from the equation $-\dfrac{dh}{dt} = C''h^{n}$, where $n = \frac{1}{2}$. But, because the value of C'' will vary from one value of h and the corresponding value of $-\dfrac{dh}{dt}$ to another pair of these values (because there is error in the data), we instead try to find a value of C'' that provides a good overall fit of all the data, not just a part of the data. For example, we could take a number of values of $h^{\frac{1}{2}}$ and the corresponding values of $-\dfrac{dh}{dt}$, and calculate the values of C'' for each,

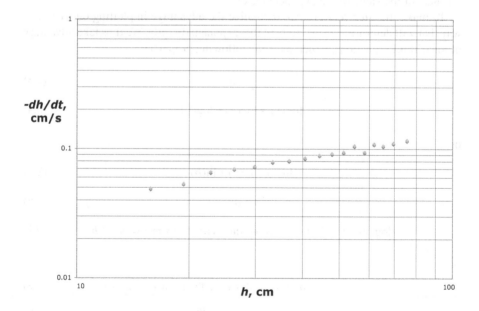

FIGURE 2.5 Dependence of tank-draining rate on liquid height. Each is plotted on logarithmic coordinates; such a plot is referred to as a log–log plot. Do you understand why we do not show the origin (0, 0) on this graph? (Because it does not exist (log 0 = -∞).)

as above, and then take the average value of C''. This would be an improvement over the first method, providing a better fit of the data as a whole.

There are still more exact and more objective ways to estimate C'' as well as to evaluate $-\dfrac{dh}{dt}$, which we will consider at a later stage.

Example 2.1 Illustration of Details of Analysis of Tank-Draining Data

Do the calculations with the data provided in this example to determine graphs like those in Figures 2.3–2.5. (The data in this example are from a separate experiment and do not quite match those shown in the figures above.)

A. Determine a form of equation to represent the data.

Solution

Using the tank-draining data shown in the plot below, we make a plot of $-\dfrac{dh}{dt}$ vs. t. From the plot below, we select seven data points (we could have chosen another number of points), and then, for each point, we draw the tangent line; then, we calculate the slope of the tangent line and recognize that this is the rise/run: $\text{slope} = \dfrac{h_{max}}{t_{max}} = \dfrac{-dh}{dt}$, where h_{max} and t_{max} are indicated approximately on the

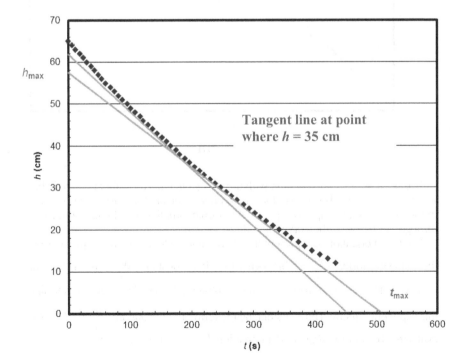

graph, and we have shown two tangent lines to represent uncertainty in the estimates (approximate upper and lower bounds).

The values determined in this way are shown in the following table:

h (cm)	h_{max} (cm)	t_{max} (s)	$-dh/dt$ (cm/s)
65	65	390	0.167
60	64.2	400	0.160
50	63	420	0.150
40	62.5	450	0.139
30	57.2	510	0.112
20	53.5	540	0.099
12	42.8	600	0.071

Thus, we have the data to make the plot of $-\dfrac{dh}{dt}$ vs. h, which is shown below. We recognize that we might represent the data with the straight line or, alternatively, with a curved line, as shown below:

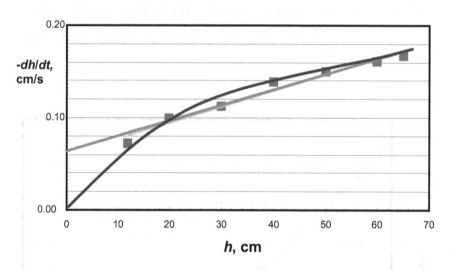

We see that the representation with the straight line fits the data rather well, but we know it is wrong, because it does not represent the limiting case corresponding to the origin on the graph; our physical reasoning implies that the curve must go through this point, as the curved line does.

We try an equation to represent the curved line sketched on the graph by using the following form: $-\dfrac{dh}{dt} = C''h^n$ (where C'' is a constant). We can narrow our estimates of the value of the exponent n by plotting the data in a way to determine this exponent directly. Thus, postulating the form of the equation $-\dfrac{dh}{dt} = C''h^n$, we evaluate n by representing our data on a log–log plot, as shown below:

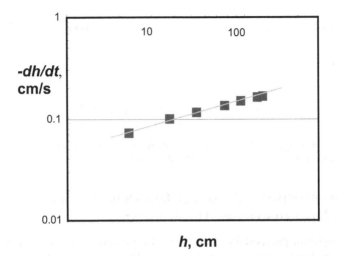

To analyze the data, we find the "best fit" straight line representing the data. To make a quick estimate of the parameters, we use two data sets:

Point 1: $h=60$ cm; $-dh/dt=0.160$ cm/s
Point 2: $h=30$ cm; $-dh/dt=0.112$ cm/s

Because this is a log–log plot, we cannot calculate the value of n directly from these numbers.

First we must take the logarithms of these numbers to obtain the slope of the straight line:

Point 1: $\log(60)=1.778$; $\log(0.160)=-0.795$
Point 2: $\log(30)=1.477$; $\log(0.112)=-0.951$

We realize that the logarithms of h and $-dh/dt$ lack meaning because these terms are not pure numbers but rather have dimensions, and that this calculation is just a convenient means to determining the slope of the line, which is the value of n; we let $y=-dh/dt$ and $x=h$:

$$\text{Slope}: n = \frac{y_2 - y_1}{x_2 - x_1} = \frac{(-0.951)-(-0.795)}{1.48 - 1.78} = 0.52$$

This slope provides a good fit with the value of n of nearly 0.5; therefore, we make the approximation that $n=\frac{1}{2}$.

Solution

B. Use the data to estimate the value of C''. To calculate the values of C'' for each point, we can simply insert the values of the data into the equation we just determined:

$$-\frac{dh}{dt} = C''h^{1/2}$$

For example, if we take the data $h = 60$ cm and $h^{1/2} = 7.746$ cm$^{1/2}$, then $\dfrac{-dh}{dt} = 0.16$ cm/s.

We get

$$C'' = \frac{\dfrac{-dh}{dt}}{h^{1/2}} = \frac{0.16\,\dfrac{\text{cm}}{\text{s}}}{7.746\,\text{cm}^{1/2}} = 0.021\,\frac{\text{cm}^{1/2}}{\text{s}}.$$

With the same approach, we can calculate the value of C'' for each point and take the average value of C''; it is equal to 0.021 cm$^{1/2}$/s.

ALTERNATIVE APPROACH TO ANALYSIS OF TANK-DRAINING DATA: EMPIRICAL APPROACH TO FITTING HEIGHT-TIME DATA

The development presented above is based on a combination of empiricism and physical reasoning about the rate of flow. The development that is presented below leads to the same result but without a starting point of considering the rate of flow directly.

Our challenge now is to use the data showing h as a function of t and to fit them empirically. The data are shown in Figure 2.1. We aim to find an equation to represent these data by using our ability to read graphs and equations. We proceed by trial and error (guess and check) or, in other words, iteratively. We begin with a simplified guess about the form of equation and, recognizing its limitations, proceed to a better guess, and then, recognizing its limitations, proceed to a still better guess, and so forth until we have a satisfactory result. Many engineering tasks are approached iteratively. We begin with an oversimplified view and proceed step by step to improve. An experienced engineer would be much more efficient than we are here.

Let us start with something very simple; in our first guess, we assume that $h = $ constant. This starting point is arbitrary, because we must start somewhere and choose something simple. We assume that the constant is the initial value of h, as shown in Figure 2.6. As we proceed with this analysis, we add comments in brackets [] to remind ourselves of the physical reasoning that we used above.

A look at the figure shows that this is not a good guess; it matches data at only one point, corresponding to $t = 0$ (where we arbitrarily made it fit). [Physically, we realize that this is not a good representation, because it means there is no flow (or $-\dfrac{dh}{dt} = 0$).]

Now, let us make a better assumption; according to our second guess, h changes with t in a linear fashion. Arbitrarily, we will make the straight-line fit good at short times, as shown in Figure 2.7.

We see some improvement, with the straight line matching some of the data well. Although for some periods, the approximation is not bad, at longer and longer times, the line undershoots the data more and more. [We also notice that the rate $-\dfrac{dh}{dt}$ does not approach zero as h approaches zero, as we know from our discussion above that it must.]

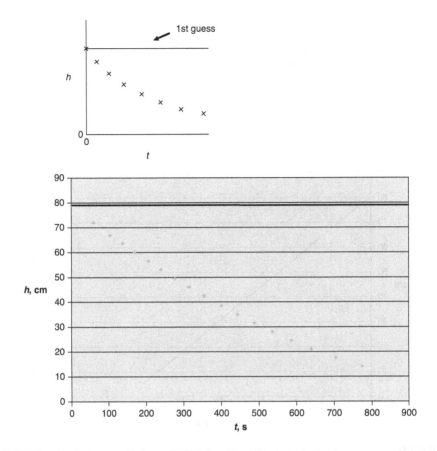

FIGURE 2.6 Attempt to fit the tank-draining data with the equation $h = $ constant; the upper plot is a sketch, the lower one shows the data.

To recap, by using calculus, we can restate our first guess as consistent with the statement that $-\dfrac{dh}{dt} = 0$; with our assumption about matching the data at $t = 0$, we overshot data except at $t = 0$. Our second guess was consistent with the statement that $-\dfrac{dh}{dt} = $ constant, which we can rewrite as $-\dfrac{dh}{dt} = \text{const} \times h^0$.

To proceed further, we make the third guess, namely, that $-\dfrac{dh}{dt} = \text{const} \times h^1$; we have just changed the power on h from 0 to 1.

To proceed with this strategy, we use calculus and define a constant as k:

$$-\frac{dh}{dt} = \text{const} \times h^1 \tag{2.9}$$

$$-\frac{dh}{dt} = kh \tag{2.10}$$

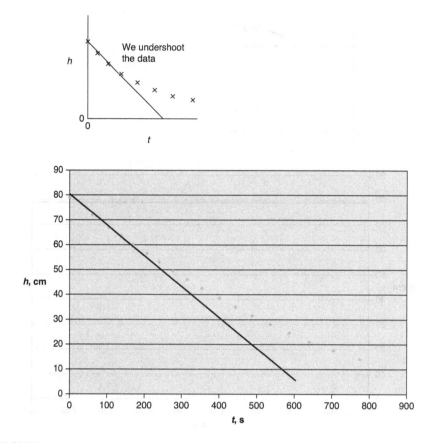

FIGURE 2.7 Representation of the data with a linear dependence of h on t. The upper plot is a sketch and is exaggerated; the lower plot shows the data.

$$-\frac{dh}{h} = -k\,dt \tag{2.11}$$

$$-\frac{dh}{h} = -d(\ln h) \tag{2.12}$$

Integrating the equation, we find

$$\ln\left(\frac{h}{h_0}\right) = -kt \tag{2.13}$$

where h_0 = the value of h at $t=0$.

By using calculus to integrate our third postulated equation, we see how to check the postulated equation against our raw data (h as a function of t). We test it by making a graph of $\ln\left(\dfrac{h}{h_0}\right)$ vs. t, recognizing that the plot will be nearly linear if our data agree well with the postulated equation. The plot is shown in Figure 2.8.

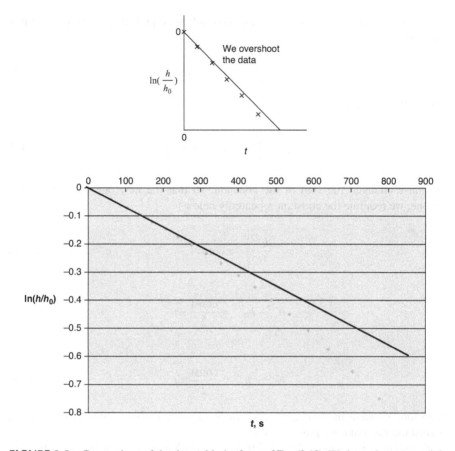

FIGURE 2.8 Comparison of the data with the form of Eq. (2.13). We have drawn a straight line to represent the data at short times. If the data had been in good agreement with Eq. (2.13) at all times, they would all have fallen near the line. Instead, we see that the line systematically overshoots the data at longer times and is not a satisfactory fit. Again, the upper plot is only a sketch.

[Note: at $t=0$, $h=h_0 \rightarrow \ln\left(\dfrac{h}{h_0}\right) = \ln(1) = 0.$]

To summarize, we see that with the second guess, $-\dfrac{dh}{dt} = \text{const} \times h^0$, we under-shoot the data, and with the third guess, $-\dfrac{dh}{dt} = \text{const} \times h^1$, we overshoot the data. Thus, it is logical to try something between these two guesses.

So our next estimate of the exponent on h is half way between 0 and 1:

$$-\frac{dh}{dt} = \text{const} \times h^{1/2} \tag{2.14}$$

Notice that this equation is familiar from our development in the preceding section; therefore, we are confident that it will give a good fit of the data.

Let us do some more calculus to understand how to plot the raw data to test this equation:

$$-\frac{dh}{h^{1/2}} = \text{const} \times dt \tag{2.15}$$

Integrating, we find

$$\int_{ho}^{h} \frac{dh}{h^{1/2}} = \int_{ho}^{h} h^{-1/2}\, dh = \text{const} \int_{0}^{t} dt \tag{2.16}$$

(We have changed the sign of the constant, so that the new one is -1 times the old one; we redefine the constant repeatedly below.)

$$\left.\frac{h^{1/2}}{1/2}\right|_{ho}^{h} = \text{const}\, t \tag{2.17}$$

$$h^{1/2} - h_0^{1/2} = \frac{\text{const}}{2} \times t \tag{2.18}$$

or

$$h_0^{1/2} - h^{1/2} = \frac{\text{const}}{2} \times t \tag{2.19}$$

Therefore, we have another equation of the form $y = mx + b$, and to test this equation against the raw data, we plot $h^{1/2}$ vs. t.

The straight-line plot in Figure 2.9 shows excellent agreement between the data and the line; the model provides a good fit of the data and is successful.

Of course, from the calculus, we see this comes from the result of the development in the preceding section:

$$-\frac{dh}{dt} = \text{const} \times h^{1/2} \tag{2.20}$$

We saw before that this equation fits the differentiated data very well, so it is no surprise that it fits the raw (undifferentiated) data well.

By using results of physics that are beyond the scope of what we are doing here, we can see how to extract further information from the tank-draining data. The result of physics that helps us is the following (which expresses the volume flow rate):

$$A\frac{dh}{dt} = C_D A_0 \sqrt{2gh} \tag{2.21}$$

where A is the cross-sectional area of the cylindrical tank and A_0 is the cross-sectional area of the orifice. It has been shown that for our orifice, the approximate value

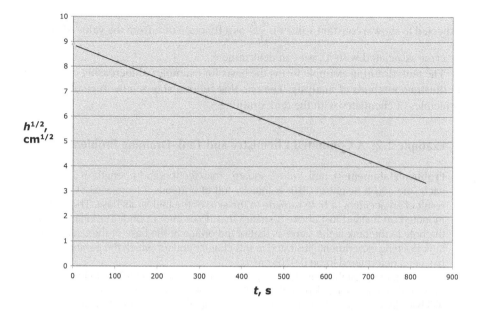

FIGURE 2.9 The data are well represented by a plot of $h^{1/2}$ vs. t, which is nearly linear.

of C_D (called the orifice coefficient) is 0.7; g is the gravitational acceleration. With this result, we can now evaluate the constant in Eq. (2.20), and see how, with the values of A and A_o, we can determine the value of g from our data.

Let us go a bit further with the analysis and practice some calculus: Eq. (2.19), upon rearrangement, gives us an expression for h as a function of t, and when we incorporate the value of the constant that we have just learned from theory, we get the following:

$$h^{1/2} = h_0^{1/2} - \frac{C_D A_o \sqrt{2\sqrt{g}} t}{2A} \tag{2.22}$$

We can square both sides of this equation to find how h depends on t. Then, we take the first derivative of each side of the equation with respect to t to find the equation for $\frac{dh}{dt}$ as a function of t, and then we differentiate again to find the following equation (can you check the calculus?):

$$\frac{d^2 h}{dt^2} = 2 \left[\frac{C_D A_o \sqrt{2} \sqrt{g}}{2A} \right]^2 \tag{2.23}$$

We thus see that this second derivative is a constant—this represents an acceleration—due to gravity. We can determine its value by graphically differentiating the data showing $\frac{dh}{dt}$ as a function of t and plotting them as a function of t. The data are expected to show a constant value of $\dfrac{d^2h}{dt^2}$ as a function of t; from the value (there will be error associated with it), we can determine g.

The tank-draining example forms the basis for examples of increasing complexity to illustrate methods of analysis. The one that follows shows how to combine some principles of chemistry with the tank draining.

Example 2.2: A Combined Chemistry and Tank-Draining Problem

Problem statement: A tank with a square cross-section and a length of 20 m on each side and a height of 20 m is initially full of aqueous 0.1-molar H_2SO_4. As a result of an accident, a hole is made in the side of the tank at its base. The tank is mounted on a stand just above a stream that flows into a pond. The diameter of the hole in the tank is the same as that of the orifice at the base of the tank in the tank-draining experiment illustrated in this chapter and for which the parameter C'' has been determined to be 0.0205 cm$^{1/2}$/s.

A. Assuming that the fluid flow properties of the 0.1-molar H_2SO_4 solution are the same as those of water, calculate the number of moles and the mass of the H_2SO_4 that will have leaked out of the tank and potentially into the stream after 1, 5, and 10 h.

Solution of Part A

Because the diameter of the hole in the tank is the same as that of the orifice at the base of the tank used to drain water as described in this chapter, we infer that the volume flow rates $(-dV/dt)$ are the same for both tanks for a given value of h. Thus, we use the results of the tank-draining experiment written for $-dV/dt$, which is $-A(dh/dt)$, where A is the cross-sectional area of the tank. We will thus use the value of the constant C'' in the equation written for $-dV/dt$ and realize that we can use it for the other tank when we use the correct values of A and h_0 for that tank. Thus, for a given fluid and a given orifice size, the proportionality constant C'' determined in one experiment with a particular tank can be used to predict the volume flow rate from a tank with a different geometry.

The cross-sectional areas of the two tanks are as follows: For the draining tank used to determine the data represented in this chapter: Area$_1 = 1,662$ cm^2; and for the cross-sectional area of the tank containing the H_2SO_4: Area$_2 = 20$ m $\times$ 20 m $= 400$ m$^2 = 4.00 \times 10^6$ cm^2.

From the development in this chapter, we have the result that $\dfrac{-dh}{dt} = C''h^{1/2}$, and we need to use the value of C'' to account for the volume flow rate as a function of t. The known volume flow rate is $\dfrac{-dV_1}{dt} = \text{Area}_1 \times \dfrac{-dh}{dt} = \text{Area}_1 \times C'' \times h^{1/2}$.

We use this information to describe the draining from the tank containing H_2SO_4, realizing that the value of h_0 is different from that of the tank used to drain water. Because the volume flow rate for each tank is the same at a given value of h:

$$\text{Area}_2 \frac{-dh}{dt} = \frac{-dV_2}{dt} = \frac{-dV_1}{dt} = \text{Area}_1 \times C'' \times h^{1/2}$$

With the equation for $-dV/dt$ for this problem, we can integrate as was done before to find V as a function of t:

rearrange: $\dfrac{-dh}{h^{1/2}} = \dfrac{\text{Area}_1}{\text{Area}_2} C''dt = C'dt$ and $C' = \dfrac{\text{Area}_1}{\text{Area}_2} C''$

$$C' = \frac{1{,}662 \text{ cm}^2}{4.00 \times 10^6 \text{ cm}^2} \times 0.0205 \frac{\text{cm}^{1/2}}{\text{s}} = 8.52 \cdot 10^{-6} \frac{\text{cm}^{1/2}}{\text{s}}$$

Then the integral $\displaystyle\int \frac{-dh}{h^{1/2}} = \int C'dt$.

We get:

$$2h^{1/2} - 2h_0^{1/2} = -C't, \ h = \left(-\frac{1}{2}C't + h_0^{1/2}\right)^2$$

The volume of fluid lost from the tank in a given time is the volume of fluid that has flowed into the stream in that time. Thus, we can determine h after certain time by applying the equation $\Delta V = \text{Area}_2 \times \Delta h = \text{Area}_2(h_0 - h)$.

Number of mols: $\Delta(\text{Number of mols}) = \Delta V \times \text{Concentration}_{H_2SO_4}$

Mass of H_2SO_4: $\Delta\text{Mass} = \Delta(\text{Number of mols}) \times \text{Molecular Weight}_{H_2SO_4}$
And so we determine the number of mols and mass of H_2SO_4 as requested:
Known:

$\text{Concentration}_{H_2SO_4} = 0.1\dfrac{\text{mol}}{\text{L}} = 100 \text{ mol/m}^3$, $\text{Molecular Weight}_{H_2SO_4} = 98\dfrac{\text{g}}{\text{mol}}$

Time: when $t=1$ $h=3.6\times10^3$ s

$$h = \left(-\frac{1}{2}C't + h_0^{1/2}\right)^2$$

$$h = \left(-1/2 \times 8.52\times10^{-6} \frac{\text{cm}^{1/2}}{\text{s}} \times 3{,}600 \text{ s} + (20 \text{ m})^{1/2}\right)^2 = 1{,}998.6 \text{ cm} = 19.986 \text{ m}$$

$$\Delta V = \text{Area}_2 \times \Delta h = 400 \text{ m}^2 (20 \text{ m} - 19.986 \text{ m}) = 5.6 \text{ m}^3$$

The change in the number of mols is

$$\Delta(\text{Number of mols}) = \Delta V \times \text{Concentration}_{H_2SO_4} = 5.6 \text{ m}^3 \times 100 \text{ mol/m}^3 = 560 \text{ mol}$$

Mass of H_2SO_4:

$$\Delta\text{Mass} = \Delta(\text{Number of Mols}) \times \text{Molecular Weight}_{H_2SO_4}$$

$$= 560 \text{ mol} \times 98\frac{\text{g}}{\text{mol}}$$

$$= 54{,}880 \text{ g} = 54.9 \text{ kg}$$

Next, when $t = 5\,h = 1.8 \times 10^4\,s$,

$$h = \left(-1/2 C't + h_0^{1/2}\right)^2$$

$$h = \left(-1/2 \times 8.52 \times 10^{-6} \frac{cm^{1/2}}{s} \times 1.8 \times 10^4\,s + (20\,m)^{1/2}\right)^2 = 19.931\,m$$

$$\Delta V = \text{Area}_2 \times \Delta h = 400\,m^2 (20\,m - 19.931\,m) = 27.4\,m^3$$

Number of mols:

$$\Delta(\text{Number of mols}) = \Delta V \times \text{concentration}_{H_2SO_4} = 27.4\,m^3 \times 100 \frac{mol}{m^3} = 2.74 \times 10^3\,mol$$

Mass of H_2SO_4:

$$\Delta \text{Mass} = \Delta(\text{Number of mols}) \times (\text{Molecular weight})_{H_2SO_4}$$

$$= 2.74 \times 10^3\,mol \times 98 \frac{g}{mol} = 268.6\,kg$$

When $t = 10$, $h = 3.6 \times 10^4 s$,

$$h = \left(-1/2 \cdot C' \cdot t + h_0^{1/2}\right)^2$$

$$h = \left(-1/2 \times 8.52 \times 10^{-6} \frac{cm^{1/2}}{s} \times 3.6 \times 10^4\,s + (20\,m)^{1/2}\right)^2 = 19.86\,m$$

$$\Delta V = \text{Area}_2 \times \Delta h = 400\,m^2 (20\,m - 19.86\,m) = 54.76\,m^3$$

Number of mols:

$$\Delta(\text{Number of mols}) = \Delta V \times \text{Concentration}_{H_2SO_4} = 54.76 \times m^3 \times 100 \frac{mol}{m^3}$$

$$= 5.48 \times 10^3\,mol$$

Mass of H_2SO_4:

$$\Delta \text{Mass} = \Delta(\text{Number of mols}) \times (\text{Molecular weight})_{H_2SO_4}$$

$$= 5.48 \times 10^3\,mol \times 98 \frac{g}{mol} = 537\,kg$$

B. Assuming that all the flow of fluid out of the pond can be stopped immediately after the tank begins to leak and that the hole in the tank can be plugged after 4 h, calculate the mass of sodium bicarbonate that should be added to the pond to neutralize the acid.

Solution of Part B

We use the same approach as in Part A to determine the number of mols in the pond after the flow has proceeded for 4 h, assuming that all the H_2SO_4 that has leaked from the tank during that period ultimately flows to the pond. The volume of fluid lost from the tank in a given time is the volume of fluid that has flowed into the stream and ultimately into the pond during the leak. Using the total time of 4 h, we determine the change in height and thus the volume of fluid, and then with a knowledge of the chemistry, vused to neutralize the acid.

Time: when $t = 4$ h $= 1.44 \times 10^4$ s,

$$h = \left(-1/2 C't + h_0^{1/2}\right)^2$$

Height after 4 h:

$$h = \left(-1/2 \times 8.52 \times 10^{-6} \frac{cm^{1/2}}{s} \times 1.44 \times 10^4 \, s + (20 \, m)^{1/2}\right)^2 = 19.95 \, m$$

The volume change: $\Delta V = \text{Area}_2 \times \Delta h = 400 \, m^2 (20 \, m - 19.95 \, m) = 21.93 \, m^3$

From above, number of mols of H_2SO_4:

$$\Delta(\text{Number of mols}) = \Delta V \times \text{Concentration}_{H_2SO_4} = 21.93 \, m^3 \times 100 \, \frac{mol}{m^3}$$

$$= 2.193 \times 10^3 \, mol$$

For the neutralization reaction, we have the stoichiometry:

$$2NaHCO_3 + H_2SO_4 = Na_2SO_4 + 2CO_2 + 2H_2O$$

So, to neutralize the acid in the pond, we need twice the number of mols of sodium bicarbonate as of the acid that flowed into the pond:

$$\Delta(\text{Number of mols})_{NaHCO_3} = \Delta(\text{Number of mols})_{H_2SO_4} \times 2$$

$$= 2.19 \times 10^3 \, mol \times 2$$

$$= 4.39 \times 10^3 \, mol$$

Also, we know:

$$\text{MolecularWeight}_{NaHCO_3} = 84 \frac{g}{mol}$$

And from this, mass of $NaHCO_3$ is:

$$\Delta Mass = \Delta(\text{Number of Mols})_{NaHCO_3} \times (\text{Molecular Weight})_{NaHCO_3}$$

$$= 4.39 \times 10^3 \text{ mol} \times 84 \frac{g}{mol} = 368 \text{ kg}$$

In sum, we need 368 kg of sodium bicarbonate.

The next example shows how we can predict the simultaneous filling and draining of a tank when the water flowing into the tank is the water draining from a tank above it.

Example 2.3: Using Tank-Draining Data to Predict Liquid Level in a Tank That Is Simultaneously Being Filled and Drained

Problem statement: Consider two tanks that are almost identical to the one used to generate the data presented earlier in this chapter, except for the orifice diameter; the fluid in them is water. The tanks have barriers (lips) above the outlets that prevent draining to a value of h less than about 10 cm. Assume that one tank is directly above the other so that it drains into it and that the initial water level in the upper tank is 70 cm and that the initial water level in the lower tank is 40 cm. At time $t = 0$, the plug in the drain of each tank is removed, so that water starts flowing from the upper tank into the lower one, and simultaneously water starts flowing out of the lower tank to the drain. Use the data in Table 2.1 (column 2) for draining of a single tank that is identical to the two tanks and prepare a graph showing the approximate height of water in each tank as a function of t. Determine points on the graph *approximately* by the following method: Use the table of the data showing the water level (height above the orifice, h) as a function of t (these data are shown in the second column of the table). Consider a time increment of 715 s and use the data to find out how much the water level in the top tank falls in this increment; this result determines how much the water level in the lower tank would have increased in the same interval, provided that the lower tank was not also losing water at the same time. But the lower tank was draining, so we also need to calculate the fall in the water level from the lower tank, but we will do this only approximately by neglecting the increment of inflow and using the graph of data and the method stated above. Then we determine an approximate value for the change in the water level in the lower tank by combining the estimate of the inflow and outflow terms and proceed analogously for the next time increment, and so on until the tanks are close to being empty. Make a graph showing h in each tank as a function of t. Assume that at $t = 0$ the liquid height in the upper tank is 70 cm and that in the lower tank is 40 cm.

Solution

The data characterizing the upper tank, the one that is simply draining, are very similar to those shown in Figure 2.1b, except for the different orifice size. Proceeding from the top of Table 2.1 toward the bottom, we see the value of h falling as time increases. The third column from the right shows the data that illustrate the draining of the upper tank; for example, in the first increment of 2543 − 1738 s, the water level fell 4.0 cm ($\Delta h_1 = -4.0$ cm; this value is negative because the water level fell). Thus, during this first interval, the water flowing into the lower tank

TABLE 2.1
Tank-Draining Data and Calculations for the Problem Solution

Interval Number	Height of Water at Various Times in Upper Tank (t in s, h in cm) at Start of Interval	Height of Water at Various Times in Upper Tank (t in s, h in cm) at End of Interval	Height of Water at Various Times in Lower Tank (t in s, h in cm) at Start of Interval	Height of Water at Various Times in Lower Tank (t in s, h in cm) at End of Interval	Change in Height of Water in Upper Tank in Interval, Δh_1 (cm)	Change in Height of Water in Lower Tank in Interval, Δh_2 (cm), Based on Assumption that there Had Been no Inflow (cm)	Change in Height of Water in Lower Tank in Interval, with Inflow from Upper Tank Accounted for, $-\Delta h_1 -\Delta h_2$ (cm)
1	(1,738, 70)	(2,453, 66.0)	(7,457, 40)	(8,172, 36.6)	−4.0	−3.4	+0.6
2	(2,453, 66.0)	(3,168, 61.8)	(7,318, 40.6)	(8,033, 37.2)	−4.2	−3.4	+0.8
3	(3,168, 61.8)	(3,883, 57.8)	(7,135, 41.4)	(7,850, 38.0)	−4.0	−3.4	+0.6
4	(3,883, 57.8)	(4,598, 53.9)	(6,998, 42)	(7,713, 38.6)	−3.9	−3.4	+0.5
5	(4,598, 53.9)	(5,313, 50.1)	(6,885, 42.5)	(7,600, 39.1)	−3.8	−3.4	+0.4
6	(5,313, 50.1)	(6,028, 46.5)	(6,795, 42.9)	(7,510, 39.5)	−3.6	−3.4	+0.2
7	(6,028, 46.5)	(6,743, 43.1)	(6,751, 43.1)	(7,466, 39.7)	−3.4	−3.4	0.0
8	(6,743, 43.1)	(7,458, 39.8)	(6,751, 43.1)	(7,466, 39.7)	−3.3	−3.4	−0.1
9	(7,458, 39.8)	(8,173, 36.6)	(6,773, 43.0)	(7,488, 39.6)	−3.2	−3.4	−0.2
10	(8,173, 36.6)	(8,888, 33.6)	(6,818, 42.8)	(7,533, 39.4)	−3.0	−3.4	−0.4
11	(8,888, 33.6)	(9,603, 30.8)	(6,908, 42.4)	(7,623, 39.0)	−2.8	−3.4	−0.6
12	(9,603, 30.8)	(10,318, 28.1)	(7,044, 41.8)	(7,759, 38.4)	−2.7	−3.4	−0.7
13	(10,318, 28.1)	(11,033, 25.5)	(7,203, 41.1)	(7,918, 37.7)	−2.6	−3.4	−0.8
14	(11,033, 25.5)	(11,748, 23.1)	(7,388, 40.3)	(8,633, 36.9)	−2.4	−3.4	−1.0
15	(11,748, 23.1)	(12,463, 20.9)	(7,620, 39.3)	(8,335, 35.9)	−2.2	−3.4	−1.2
16	(12,463, 20.9)	(13,178, 18.8)	(7,903, 38.1)	(8,618, 34.7)	−2.1	−3.4	−1.3
17	(13,178, 18.8)	(13,893, 16.9)	(8,214, 36.8)	(8,929, 33.5)	−1.9	−3.3	−1.4

would have increased its level by 4.0 cm, provided that the lower tank had not been draining. However, it was draining, and the initial value of h was 40 cm, and so we estimate that the level in the lower tank would have fallen by 3.4 cm—the change in the height of the upper tank during the interval starting when its water level was 40 cm. Thus, the net change in the height of liquid in the lower tank is estimated to be $4.0 - 3.4 = 0.6$ cm. This estimate is not exact because it neglects the effect of the flow rate out of the lower tank resulting from the water added from the upper tank during the time increment. But because the water that flowed into the lower tank during this time increment was a small fraction of the total amount of water in the lower tank (see how this statement is normalized), it was a good estimate to neglect the inflow during this increment.

If we make the increment smaller, we make the estimates better. In the limit as the increment approaches zero, the estimated values of h in each tank approach the exact values.

The results of the calculation are summarized in Table 2.1 and Figure 2.10. We emphasize that the height of water in the lower tank initially increases, because the water level in the upper tank is initially greater than that in the lower tank, so that the inflow to the lower tank is initially greater than the outflow. When the water level in the lower tank becomes equal to that in the upper tank, the rate of flow into the lower tank equals the rate of flow out of that tank, so that the value of the height of the liquid in the lower tank reaches a maximum.

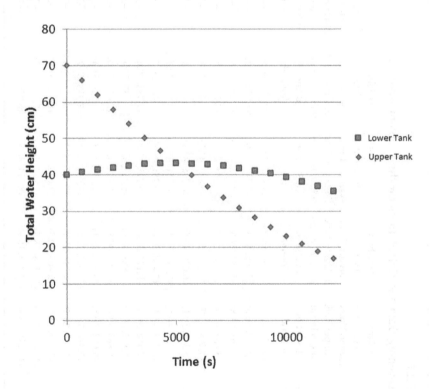

FIGURE 2.10 Results of calculations taken from Table 2.1 showing how as the liquid level in the upper tank falls, that in the lower tank receiving the fluid from the upper tank first increases and then decreases.

Essentially all we have done in solving the problem of Example 2.3 is to read a graph or table of data; this example is an introduction to what we call numerical methods of analysis, methods with which we make approximations that can be as exact as we wish. We make them more and more exact by doing more and more fine-grained calculations.

We realize that the problem can also be solved with equations—can you set up the equations that describe the heights of liquid in the two tanks as a function of time and begin to solve the problem that way?

Further Analysis of Tank-Draining Data: Effect of Orifice Diameter on Flow Rate

There is a variable that we expect to affect the flow rate of water from the tank that we have not yet investigated—the diameter of the orifice at the tank base. The above-stated theoretical result shows that the rate increases in proportion to the orifice area. To determine this dependence, experiments were done with orifices of various diameters mounted on a tank. Data such as those shown above were used to determine the flow rate of water at a particular value of h. The data are plotted in Figure 2.11. They confirm the expected linear dependence of flow rate on orifice area.

What limiting case can you identify in the plot shown in Figure 2.11? Do the data extrapolated to this limiting case match your expectations? How does your reasoning relate to that stated earlier in this chapter regarding the limiting case as the value of h approaches zero? How well do the data shown in Figure 2.11 agree with the theoretical result stated in Eq. (2.22)? Do you understand how values of $-dh/dt$ were determined at a value of $h = 31.5$ cm? What would you

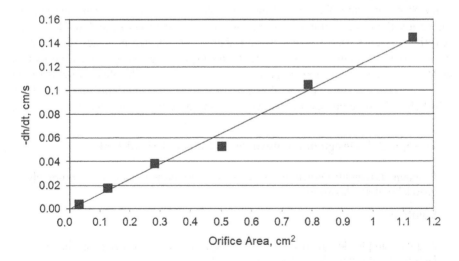

FIGURE 2.11 Dependence of $-dh/dt$ (which is proportional to the flow rate of water from the tank) on cross-sectional area of orifice mounted at the base. The data were determined at a value of the liquid height above the orifice (h) of 31.5 cm. The data are well represented by a straight line that passes through the origin.

expect to be the form of the plot determined at another value of h? How would you predict the graph of $-dh/dt$ at any given value of h?

DESIGN OF TANKS FOR CONSTANT FLOW RATES

Let us think about a design of draining tanks to provide a constant flow rate. To understand the basis for the design, consider a tank like the one for which we presented data in this chapter. The flow rate of water from the tank decreased as the water level fell, because the force acting of the water and causing the flow from the tank—the force of gravity acting on the water—decreased as the *height* of the water above the orifice decreased. If we wanted the flow rate of water from the orifice to remain constant, we could, for example, keep the tank full at all times by having water flow into the tank at such a high rate that the tank was always full, with the excess water that did not pass through the orifice flowing out of the tank over the top rim. This design would provide a constant height of water in the tank.

There is another way to construct a tank to give constant water flow rates for part of the draining period—without needing an excess of flowing water. We can do this by sealing the top of the tank with a cover and inserting through the cover—a seal—a tube that extends vertically toward the base of the tank. As this tank drains, the water that leaves the tank is replaced by air entering the tank through the tube; bubbles of air rise from the tube tip to the top of the tank as the water level falls. This design provides a constant flow rate of water as long as the water level is above the level of the tube tip.

We can understand this behavior by realizing what the pressures are at various positions in the tank. When the tank is open, the pressure at the top of the liquid is atmospheric pressure. The pressure at the downstream edge of the orifice is also atmospheric pressure. We see that the atmospheric pressure is not causing the flow; rather, it is the force of gravity working on the water that causes the flow. Now, if we seal the tank and insert the tube as described above and let the water start to flow, then the pressure at the tube tip (where the bubbles emerge) is atmospheric pressure, and the force driving the flow of the water corresponds to the height of water between the orifice and the tube tip. The flow rate of water is constant until the water level falls below the tube tip, and then it declines—because the "head" declines.

Example 2.4 Design of an Inexpensive Constant-Head Tank

Problem statement: Design an inexpensive constant-head tank from commonly used laboratory glassware.

Solution

A good, simple design made from standard laboratory glassware is shown in Figure 2.12; it is a modified glass separatory funnel.

This is a separatory funnel with a rubber stopper in the opening at the top to provide a seal; a hole is drilled in the stopper, and a glass tube is fitted tightly into it so that the seal is preserved. The valve at the bottom of the separatory funnel is a

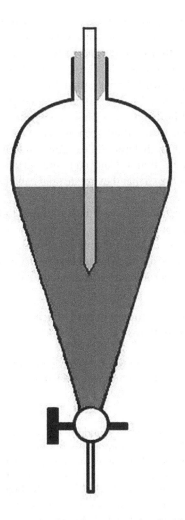

FIGURE 2.12 Schematic representation of a constant-head tank.

stopcock. Regulating the position of the stopcock provides a control over the flow rate; so does the height of the tip of the tube.

MEASUREMENT OF FLUID FLOW RATES

How do engineers measure flow rates? There are many devices available for such measurements, ranging from the simple and inexpensive to those with electronic components and accompanying equipment to continuously display, record, and transmit data. A simple "bucket and stopwatch" method is common, whereby one weighs a bucket (determines the tare weight), uses it to accumulate all the liquid in a stream for a measured time, and reweighs it, determining the average flow rate for the measured time.

Continuous readings of liquid flow rates are made with flow meters. An example is a rotameter: in this device, liquid flows upward through a vertically mounted tube, typically made of glass with a slightly tapered inside diameter that increases from bottom to top. In the tube is a solid "float" (or "bob") that is pushed up by the force of the upward-flowing liquid while being pulled down by gravity. The forces balance at a steady state (constant liquid flow rate), and the height of the float in the tube, which has a scale etched in the glass or plastic, determines the flow rate. The flow rate depends on properties of the fluid such as density and viscosity, and floats of different mass can be used for different fluids. Can you figure out why the float must have a higher density than the liquid? Can you figure out why some floats are grooved and colored (and transparent), so that they rotate (and can be observed to rotate) when the fluid flows?

Any device such as a rotameter has to be calibrated to provide quantitative and reliable data. The calibration for a liquid could be done with the bucket and stopwatch method, with measurement of the volume of fluid flowing in a measured time at each of a number of float positions. The calibration depends on the fluid properties. A method for calibration of a rotameter is indicated by the next example.

Example 2.5 Design of an Inexpensive Device for Measuring Gas Flow Rates

Problem statement: Design an inexpensive device for measurement of flow rates of gases at low pressures from commonly used laboratory glassware.

Solution

A good, simple design made from standard laboratory glassware is shown in Figure 2.13; it is a modified glass burette. A glass blower modified the burette to allow inflow of gas from the side to allow mounting at the base of the rubber bulb, to which an aqueous solution of soap or detergent is added (some commercial dishwashing detergent solutions work well). When the bulb is squeezed, it creates soap films that are intercepted by the gas stream entering from the left. The gas stream pushes the film up the cylinder. A stopwatch is used to measure the rate of rise of the films and thus the volume flow rate of the gas.

Can you figure out why the device stops working well when the inside walls of the cylinder dry out?

GAS-SOLID AND LIQUID-SOLID FLOW AND SEPARATION

When particles of a solid are placed in a tube and fluid flows upward through the tube at a low rate, the fluid simply flows around the particles—through the spaces between them. We call this a fixed bed of particles. It allows good contacting of the fluid and the solid, for example for a chemical reaction involving them both.

When the flow rate of fluid increases sufficiently, it exerts enough force to cause the bed of particles to expand, as shown schematically in Figure 2.14. The bed is then said to be fluidized. Quite complex flow patterns emerge in fluidized beds, depending on the particle size distribution, the particle density, and the physical properties of

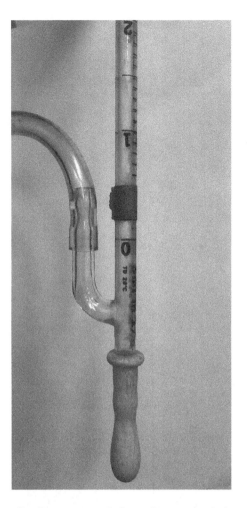

FIGURE 2.13   A soap-film flow meter made from a burette. A solution of detergent is placed in the rubber bulb at the base, and when this is squeezed, the level of the solution rises to the level of the gas inlet at the left, and the flowing gas creates films that rise in the tube, with the rate of rise determining the gas flow rate.

the fluid; sometimes bubbles form (Figure 2.14). Further increases in the fluid flow rate lead to carry-over of the particles. The particles are then said to be entrained. Thus, we see a way to transport solid particles. We also see that we have a basis for separating smaller from larger particles or less-dense particles, because the particles that are smaller and less dense are entrained at lower fluid flow rates than the larger and denser particles. Thus, by increasing the fluid flow rate step by step and collecting the particles that flow out of the tube, we can collect particles that are fractionated according to size or density.

Another way of separating particles into various size ranges simply involves passing them through standard sieves; these are available in stacked trays, and shaking them leads to separation of the particles by size. For example, in the U.S. Standard

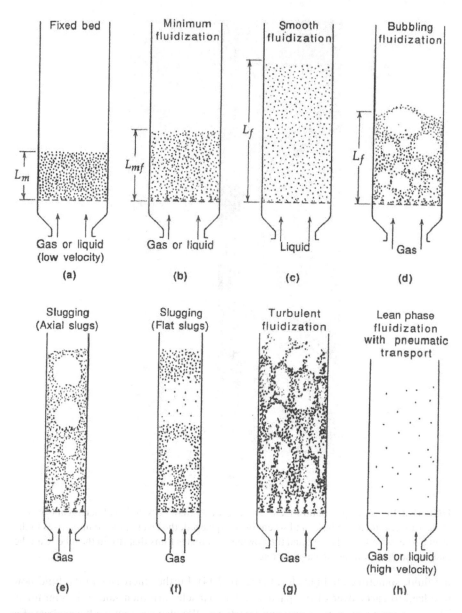

FIGURE 2.14   Schematic representation of gas or liquid flow through fixed and fluidized beds, with definitions of terms illustrated by the drawings. The symbol L refers to the bed height, with L_m referring to the minimum value and L_{mf} referring to the value at the gas or liquid flow rate at which fluidization starts. From Klinzing, G. E., Marcus, R. D., Rizk, F., Leung, L. S. (1997). Flow regimes in vertical and horizontal conveying. In: Pneumatic Conveying of Solids. Powder Technology Series, vol. 8. Springer, Dordrecht. Reproduced with permission.

Sieve Series for use in laboratories, sieve number 170 is a cylinder with a diameter of about 20 cm and a height of about 7 cm, with a screen mounted in its base that has square openings with the openings having an edge dimension of 0.090 mm. These trays are made of brass or steel, and old ones are regarded as collectables.

Some particles are so small that they are readily transported in the atmosphere; we see them as dust particles. These are important to human health. Particles of smoke formed in forest fires are highly toxic, and other components of dust are toxic too.

Other components floating in the atmosphere are of concern too, including aerosols—this term includes both small solid particles and liquid droplets. Many of these originate from humans and other animals, including small droplets from animal breath. These transmit diseases. Some emerge from ocean spray and from volcanos. There are many other sources. Aerosols are deposited in the lungs and make their way into human bodies.

Aerosols are removed from air by filtration. High-efficiency particulate air (HEPA) filters can remove high percentages of them and are important in managing indoor air quality.

Pollutants in the air combine with aerosols, which transport them. Many toxic chemicals are present in the air we breathe and in the water in our lakes and streams.

Example 2.6 Design of an Inexpensive Air Purification Device

Problem statement: Design a device for cleaning indoor air made from inexpensive and readily available components.

Solution

An excellent example of engineering design is the Corsi-Rosenthal box, which emerged, and gained widespread attention, in the preceding few years, as people reacted to the COVID-19 virus. The cube-shaped device is made from a box fan with a square cross-section and air filters that are used in home air-handling systems and are rated as MERV (minimum efficiency reporting value) rather than the more expensive HEPA (Figure 2.15). The device is held together by duct tape. The components cost about $60.

Recap and Review Questions

We have now carried through an empirical analysis of the draining tank, using experimental results to determine rates of the tank draining as a function of the height of liquid above the base and comparing the form of the height–time graph with forms of familiar equations as a basis for selecting an equation to represent the data. In making the comparison, we were guided by physical reasoning, realizing that as the water level declined, the rate of draining declined and as the water level approached zero, the rate of draining approached zero. We used a separate approach to do analysis of the same data, making a graph of the liquid height as a function of time and proceeding to compare the form of the graph with those of familiar functions.

We used the draining tank as a starting point to introduce flow in simple devices, leading to methods for measurement of fluid flow rates. The idea of calibration is essential here, extending to many engineering measurements besides flow rate measurements.

FIGURE 2.15 Photograph of inexpensive air filtration device. *Source*: Reproduced with permission from R. Dal Porto, M. N. Kunz, T. Pistochini, R. L. Corsi, and C. D. Cappa, Chacterizing the performance of a do-it-yourself (DIY) box fan air filter, *Aerosol Sci. Technol.*, **2022**, *56*, 564–572, Supplemental Information.

Test your understanding of the terms used here and in the preceding chapter. Is the tank draining a transient process? How could you modify the equipment or the operation and make it a steady-state process? What does the term "rate" mean in the draining tank example? How is rate different from average rate? Test your understanding of calculus by checking whether all the steps in our derivations that involve calculus have been done correctly.

PROBLEMS

2.1. With a graphical method, use data you have collected for the draining tank or data presented in Table 2.1 to determine values of d^2h/dt^2. Plot these data as a function of h. Determine an equation for d^2h/dt^2 as a function of h. What do you learn about the errors in differentiated (and twice differentiated) data?

2.2. Consider two tanks that are identical to the one used to determine the data shown in Figure 2.1b; the fluid in them is water. Assume that one tank is directly above the other and that the initial water level in the upper tank is 70 cm and that in the lower tank is 35 cm. At time $t = 0$, the plug in the drain of each tank is removed, so that water starts flowing from the upper tank into the lower one, and simultaneously water starts flowing out of the lower tank to the drain. Make a graph showing the *approximate* height of water in each tank as a function of t. Determine points on the graph *approximately* by the following method: Use a graph of your data showing the water level (height above the orifice) as a function of t. Consider a time increment of 100 s and use your data to find out how much the water level in the top tank falls in this increment; this result tells you how much the water level would have increased in the lower tank in the same interval, provided that the lower tank was not also losing water at the same time. But the lower tank was draining, so calculate the fall in the water level from the lower tank *neglecting the inflow* by using your graph of data and the method stated above. Then get an approximate value for the change in the water level in the lower tank by combining the estimate of the inflow and outflow terms (did the water level rise or fall during this increment?). This calculation gives you an approximate liquid level in the lower tank after the interval of 100 s. Now proceed to do the same kind of calculation for the next time increment of 100 s, and so on until the tanks are almost empty.

2.3. Consider four tanks that are identical to the one used to determine the data shown in Figure 2.1b; the fluid in them is water. Assume that the tanks are mounted in three levels, the top tank (tank 1) at level one, the next two tanks (tanks 2 and 3) at level two, below level one, and the bottom tank (tank 4) at level three, below level 2. When the drain of tank 1 is open, the fluid in it flows directly into tank 2. When the drain of tank 2 is open, the fluid in it flows directly into tank 4. When the drain of tank 3 is open, the fluid in it flows directly into tank 4. When the drain of tank 4 is open, the fluid in it flows directly into the sewer. Also assume the following:

tank 1 is initially filled to a height above the orifice of 70 cm;
tank 2 is initially filled to a height above the orifice of 70 cm;
tank 3 is initially filled to a height above the orifice of 70 cm;
tank 4 is initially empty;
the plugs are removed from tanks 1, 2, and 3 at time $t = 0$;
the plug is removed from tank 4 when the liquid level above the orifice is 35 cm.

A. Sketch the system of tanks and include the given dimensions.
B. For a certain period, the liquid level in one of the tanks does not change; identify the tank and explain why the liquid level is constant for this period.
C. Use a graphical method and the data that you obtained in the laboratory or data presented in this chapter to determine the approximate height of liquid in each tank above the orifice as a function of time t and show the results in a graph. The graph will have four lines; label them 1, 2, 3, and 4, respectively, for the tanks with these numbers.

2.4. Consider three tanks that are identical to the one described in this chapter. Assume that two of the tanks are mounted directly above the third so that the flow from the upper two tanks goes directly into the lower tank. Also assume the following:

the first upper tank (tank A) is initially filled to a height above the orifice of 50 cm;

the second upper tank (tank B) is initially filled to a height above the orifice of 25 cm;

the lower tank (tank C) is initially empty;

the plug is removed from each of the two upper tanks at $t = 0$;

the plug is removed from the lower tank when the liquid level above the orifice in the lower tank is 40 cm.

Use graphical methods and data shown in this chapter to determine the height of liquid in each tank above the orifice as a function of time t and show the results in a graph. The graph will have three lines; label them A, B, and C, respectively, for the two upper tanks and lower tank.

2.5. Consider a cylindrical tank like the one used to generate the tank-draining data presented in this chapter, but with modifications. The modified tank has a cap at the top, and it is sealed. The vertical distance between the base of the tank and the base of the cap is 50 cm. At the base of the tank, a vertical pipe is mounted, with a diameter of 5.0 cm, and at its base is a plate with a circular orifice having a diameter of 0.50 cm. The tank is outfitted with a vertical tube, sealed and extending through the cap and reaching the pipe entry at the base of the tank. With this design, we have a constant-head tank with a pipe at the base.

   A. Sketch the apparatus and show the dimensions.

   B. If the tank is initially filled with water, estimate the time required for it to drain until the height of water above the orifice is 1.0 cm.

2.6. Consider the oil painting *The Danaides* by John William Waterhouse and the Greek myth on which it is based. Images of the painting are accessible on the internet. Model the draining vessel on the basis of the painting, and estimate the essential dimensions from the sizes of the people in the painting. Do the necessary research regarding the myth, and estimate how much water each sister must pore into the vessel in one day. Propose a hellacious work schedule for the sisters. State your assumptions and approximations and the basis for each; be sure that your research determines the number of sisters.

2.7. A fast-flowing stream provides excess water for driving a waterwheel. It is desired to drive the waterwheel at a constant rate and to return all the water to the stream. Provide the sketch of a low-cost design to meet the goal, and state specifically how it could provide a water flow rate of 1.50 m³/min.

2.8. A stream is fed by melting snow. The stream flows into a reservoir that was initially empty, and there was no flow out of it (all the water was collected in the reservoir). The total volume of water in the reservoir was measured as a function of time. The data in the table below show the volume of the water (V) that flowed into the reservoir as a function of the time (t).

A. Estimate the volume of water that had flowed into the reservoir after 7 days (that is, at $t=7$ days).

B. Estimate the volume flow rate after 3 days (that is, at $t=3$ days). Recall that the volume flow rate is dV/dt.

C. Explain why you know that the relationship between the volume flow rate and time is not linear.

D. Propose an equation to represent how the volume flow rate of water depends on the time (t). Explain briefly why your equation is a good choice to represent the data in the table. (Hint: look carefully at the data and consider how much change of volume occurs when time changes by a factor of two.)

$t=$ time, days	V, volume of water collected in reservoir (ft³)
0	0
2	500
4	2,000
6	4,500
8	8,000
10	12,500

2.9. The data in the table below were found experimentally. They show the rate of a chemical reaction (r) as a function of the concentration of the reacting molecules (c).

A. Identify any limiting cases of these data. That is, identify, by circling them on a plot, with linear scales, any regions on the graph where the results are represented by an especially simple form of equation such as a linear form, $r=mc+b$. Circle only regions of the graph that are well approximated by limiting cases.

B. Write a simple equation (model) to represent the data for each limiting case. In other words, state with an equation for each limiting case how r is a function of c.

C. Consider the following possible equations to represent all the data, and choose the one that best represents all the data in the table and on the plots:

$r=b$ (where b is a constant)
$r=mc+b$ (where m and b are constants)
$r=mc^n$ (where m and n are constants)
$r=m/c$ (where m is a constant)
$r=mc/(1+bc)$ (where m and b are constants)
$r=\sin(c)$

In other words, choose this equation to represent how r is a function of c for all values of c. Estimate the values of any *parameters* in this equation. (*Parameters* are defined as terms other than r and c; m and b are examples of parameters.)

r (mol/(L . s))	c (mol/L)
0.010	0.010
0.020	0.020
0.040	0.040
0.098	0.10
0.19	0.20
0.45	0.50
0.83	1.0
2.50	5.0
3.33	10
4.0	20
4.5	50
4.8	100
4.9	200
5.0	500
5.0	1,000

2.10. Use the data of Example 2.1 to estimate the gravitational constant g. Assume that the value of the orifice coefficient was 0.7 and that the diameter of the tank was $1,662 \, cm^2$.

2.11. The side-blotched lizard, *Uta stansburiana*, in California is character-ized by three distinct forms that feature different throat colors: orange, blue, and yellow. These three have different strategies for propagation. The males with orange are dominant and polygamous, besting the more monogamous blues, which guard their mates. The orange-throated males are in turn bested by the yellows, who in some ways resemble females and sneak into the territories of the orange-throated males to mate. The sneaker strategy of the yellows is trumped by the mate-guarding blues. Thus, each of the three types is susceptible to dominance by one of the others. It has been observed that the populations of the three types of side-blotched lizard in a particular zone cycle, with each gaining ascendancy and then losing it to another type. Below is a photograph of an orange-throated side-blotched lizard:

A. Explain how this behavior pattern is suggestive of a familiar game: rock, paper, scissors.

B. Explain why the populations of the three types of the lizard are expected to cycle.

C. Suggest a form of equation that might approximately represent the pop-ulation density of each type of side-blotched lizard in a zone as a func-tion of time.

D. Prepare a sketch showing the population of each type as a function of time, without quantifying the time.

2.12. A zoo keeper's assistant is given the assignment of using plastic tubing with an inside diameter of 1/8 inch and a donated 10,000-L tank cylindrical tank with a height equal to twice its inside diameter to design a system to provide water to four terrariums that are filled with tropical plants and house chameleons, lemurs, turtles, and butterflies, respectively. The animals require continuously flowing fresh water to drink and to maintain the humidity of their surroundings. The tank is supplied by a local stream with plenty of water to keep it filled. The terrarium with the turtles requires a flow of 10 L/h, but each of the other terrariums requires a flow of only 100 mL/h. The tank can be mounted on the roof above the terrariums, with the tank base being 10 m above the top of each terrarium. Design a simple, reliable, inexpensive system to provide water to the terrariums.

2.13. Do some research and find out what a wet-test flow meter is, and explain how it works.

2.14. People lose mass during sleep. Do some research and find some data that determine rates of mass loss. Can you find data showing how the rate of mass loss depends on the humidity of the surrounding air? Can you find data that determine rates of loss of CO_2 and of water?

2.15. An engineer plans to carry out an experiment with flowing water and needs to keep the flow rate constant at a rate of 1.0 mL/min for periods of unattended operation as long as 3 h. She has access to standard chemistry lab glassware, flexible tubing, and a valve. Design the system as simply as possible and explain how the design provides for a constant rate of water flow.

2.16. Suppose you have some beach sand and want to separate it into fractions containing various sizes of particles. Sketch simple equipment for doing this separation using flowing air and a glass tube, and explain how it works. Presuming that you would have measurements of the particle sizes determined with an optical microscope for each size fraction, explain how you would determine the mass fraction of particles in each of 10 size ranges.

2.17. Find out who Evangelista Torricelli was and his connection to the analysis of the draining tank.

2.18. Under some conditions, water leaving a tank through an orifice at the base of the tank flows as a continuous stream, but, under other conditions, the flow is markedly different: the water flows drop by drop, that is, discontinuously. Do some research and find out how researchers have used strobe lights to visualize and photograph the drops and determine their sizes and the frequency of drop formation.

2.19. Do some research and find out how sulfate aerosols get into the atmosphere.

2.20. Find data that indicate the composition of
   A.   smoke from wildfires
   B.   smoke from cigarettes.

2.21. Check evidence for the hypothesis that flu virus may be transported through the air on dust particles and animal dander. See the work by A. S. Wexler and W. D. Ristenpart.

2.22. Fish caught in the oceans are sometimes contaminated by small particles of plastic. Do some research and find out how the plastic makes it way to the fish and how it is harmful to the ecology and to human health.

2.23. Rework Example 2.3 with twice the interval used in the example. Explain how and why the results differ from those determined in the example.

3 Analysis Guided by Theory
Conservation of Mass
and Conceptualization
of Processes

ROADMAP

In this chapter, we go beyond empiricism (or just fitting of data to find an equation) to determine a model. We are now guided by a general scientific principle, the conservation of mass. This is the starting point of many analyses and models in engineering. It starts with a balance equation, and its application is illustrated for cases of increasing complexity as this chapter proceeds. We build on this foundation to begin to conceptualize processes that involve combinations of operations such as flow, mixing, and separation and introduce process flow diagrams, which are schematic summaries that show how various operations are combined, with indications of flows.

INTRODUCTION TO BALANCE EQUATIONS

Our analyses to this point have been empirical and guided by physical reasoning. We can do better; we can use a more general and more powerful method. The next level—taking us beyond empiricism—is based on fundamental scientific principles. The most basic of these principles for our purposes is the *conservation of mass*. This principle states that mass (matter) is neither created nor destroyed. An equally basic principle states that energy is neither created nor destroyed, and we will use this second principle in Chapter 6.

To use the principle of conservation of mass, we proceed by stating it in the form of a *balance equation*. Many analyses in engineering begin with balance equations. These equations are accounting statements, and we introduce them with an analogy to something familiar: bank statements.

We consider a bank account with no fees and no interest. Doing an account balance, we state that

what goes in – what goes out = what is accumulated (savings).

The balance statement takes the form of an equation written for the bank account. We refer to the account as the system over which the balance equation is written. The terms in our equation are the following, with the units stated at the right:

Average rate of "flow" of money into the account $= F_{\text{\$in}}$ \$/time
Average rate of "flow" of money out of the account $= F_{\text{\$out}}$ \$/time
Change (accumulation or loss) of money in the account $= \Delta N\text{\$}/\Delta t$ \$/time

DOI: 10.1201/9781003429944-3

where N = number of dollars (the units we choose for money) and t = time. Usually, we get bank statements once per month, and so we may do the accounting each month. Thus, the "flow" of money into the account (e.g., our paycheck and other deposits) is the total amount of money deposited into the account in that one month; the "flow" of money out is our withdrawals, plus debits that are electronic transfers, plus checks written, etc. And the accumulation is our savings (which could be positive or negative). In contrast to the flow of water from our draining tank, which is continuous, the "flow" of money into and out of the account is not continuous; rather, it is sporadic or discrete, occurring on occasions when we make a deposit or a transfer, for example.

The money balance equation is the following (where the inflow term is illustrated by pay and the outflow term is illustrated by checks written):

$$F_{Sin} \quad - \quad F_{Sout} \quad = \Delta N_j / \Delta t$$

average rate	average rate of "flow" of	dollars accumulated over
of "flow" of dollars	dollars out of account	the time considered
into account (pay)	(checks written)	(change in the number of dollars
		in account in time interval)

$$\frac{\$}{time} \qquad \frac{\$}{time} \qquad \frac{\$}{time}$$

According to this statement, *money is conserved*: the number of dollars accumulated equals the number of dollars paid into the account minus the number of dollars paid out of it. This conservation statement is true only if there are no fees or interest.

However, by analogy with our tank-draining analysis stated earlier, we could refer to the average of the flow of money into the account as the amount of money that was paid in divided by the time period in which the payments occurred—this would, for example, be the average rate of pay, say, in dollars per month. We could make similar statements about the outflow terms, say the rent in dollars per month. We reemphasize: there is no implication that these money flows proceed at constant rates; they generally do not.

CONSERVATION OF MASS APPLIED TO A TANK WITH INFLOW AND OUTFLOW STREAMS

Now, we do something analogous for a continuous flow system. What is conserved is mass. The statement of the conservation of mass is the following: what flows in (mass) – what flows out (mass) = accumulation (of mass). It does not matter what the compositions of the streams or the tank contents are. The accumulation may be positive or negative.

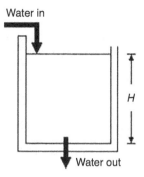

Water in

H

Water out

FIGURE 3.1 A system, consisting of a tank, characterized by a continuous inflow and outflow of water. We use this system to illustrate the principle of conservation of mass. We represent the height of liquid above the base of the tank as h, a variable, and the maximum value of h as H.

Consider a system slightly more complex than the draining tank considered in the preceding chapter; now we do the analysis for a tank with an inflow stream as well as an outflow stream, as illustrated in Figure 3.1.

Because mass is conserved,

$$\begin{matrix} \text{rate of water flow} & & \text{rate of water flow} & & \text{rate of water accumulation} \\ \text{into tank} & - & \text{out of tank} & = & \text{in tank} \\ \frac{\text{mass}}{\text{time}} & & \frac{\text{mass}}{\text{time}} & & \frac{\text{mass}}{\text{time}} \end{matrix} \qquad (3.1)$$

We proceed assuming that the tank is cylindrical; cross-sectional area $= A$.

This is a mass balance equation (sometimes a mass balance is called a material balance). Now, we write the mass balance around the tank, which is chosen to be the system; the system is also called the *control volume* for this development. Note that we are choosing to write the equation for an interval of time Δt, and we can anticipate on the basis of our consideration of average rates and point rates in the preceding chapter that—*when the flow is continuous*—we will take the limit in which this time interval approaches zero. In other words, we will start with average rates and take the limit to consider point values of rates. In the following equations, ρ represents the density of the water (which we consider to be a constant value); q_f is the volume flow rate of water into the tank, also taken to be a constant value; and q is the volume flow rate of water out of the tank (which depends on h and therefore is a variable, not a constant value). Dimensions of ρ are mass per unit volume, and dimensions of q are volume per unit time. A represents the constant cross-sectional area of the tank (assumed to be cylindrical), and h is the height above the base (h is a variable), where the orifice is mounted.

accumulation

mass which entered tank between t and $t + \Delta t$	−	mass which left tank between t and $t + \Delta t$	=	mass in tank at time $t + \Delta t$	−	mass in tank at time t
$\rho q_f \Delta t$	−	$\rho q \Delta t$	=	$\rho A h(t + \Delta t)$	−	$\rho A h(t)$
↑		↑				
volume flow rate in		volume flow rate out		h evaluated at $t + \Delta t$		h evaluated at t

$$\text{(3.2)}$$

It is emphasized that this is a mass balance equation. Each term has units of mass, such as grams.

Dividing by ρA gives

$$\frac{1}{A}\left[q_f - q\right]\Delta t = h(t + \Delta t) - h(t) \tag{3.3}$$

or

$$\frac{h(t + \Delta t) - h(t)}{\Delta t} = \frac{1}{A}\left[q_f - q\right] \tag{3.4}$$

Now, we use calculus and take the limit as $\Delta t \to 0$:

$$\frac{dh}{dt} = \frac{1}{A}\left[q_f - q\right] \tag{3.5}$$

To cast the equation in a form representing the change in volume with respect to time, we re-multiply by ρA (which is a constant) to get the following:

$$\frac{d[\rho A h]}{dt} = \rho q_f - \rho q = \rho A \left(\frac{dh}{dt}\right) \tag{3.6}$$

Remember that q_f = volume flow rate into the tank, which is taken to be unchanged with respect to time (constant) and q = volume flow rate out of the tank, which is not constant but depends on h.

Look closely at Eq. (3.6) and recognize that it provides information about how the volume flow rate q depends on the height of liquid above the orifice (h), but it does not simply relate q to h. Instead, it relates q to dh/dt.

USE OF A CONSTITUTIVE RELATIONSHIP

To proceed with this development, we need more information. Provided that the tank and the orifice are the same as what we considered in the tank-draining analysis of the preceding chapter, we have the additional information we need, that is, an equation showing how the volume flow rate of water draining from the tank depends on h:

$$q = -A\left(\frac{dh}{dt}\right) = A C h^{1/2} \tag{3.7}$$

where C = constant, which we evaluated in our analysis, on the basis of experiment.

What we are doing as we proceed with this analysis, which started with the mass balance equation, is to use an experimental result that is characteristic of the specific system we are considering (the draining tank)—it pertains to this system but not necessarily to others. The experimental result is an example of a *constitutive relationship*. A constitutive relationship is a result determined by experiment that characterizes a specific system (such as the particular tank with the particular orifice with water flowing at a particular temperature). The principle of conservation of mass, in contrast, is general rather than specific; we often proceed in analysis by starting with the general statement (such as conservation of mass) and then combining it with a specific (that is, limited) statement (constitutive relationship).

By combining the mass balance equation with the constitutive relationship for the tank, we determine the following:

$$\rho A\left(\frac{dh}{dt}\right) = \rho q_f - \rho A C h^{1/2} \tag{3.8}$$

or

$$\frac{dh}{dt} + C h^{1/2} = \frac{q_f}{A} \tag{3.9}$$

Most of us do not yet know how to solve this kind of equation (called a differential equation because of the derivative in it), but let us consider a special case, the steady state, to put ourselves on more familiar mathematical ground.

From the definition of a steady state (Chapter 1), we realize that, in this state, h does not change with respect to time; in other words, the rate of change of h with respect to time must be zero:

$$\frac{dh}{dt} = 0 \tag{3.10}$$

In this case, Eq. (3.9) simplifies to

$$C h^{1/2} = \frac{q_f}{A} \tag{3.11}$$

If, for example, we want to use this result to consider a case in which H is the maximum height to which the tank can be filled and postulate that we want the water level in the tank to be half of this value, then

$$h = \frac{1}{2}H \text{ and } q_f = \left(\frac{CA}{\sqrt{2}}\right)H^{1/2} \tag{3.12}$$

This result determines the flow rate into the tank that would be needed to keep the level of water in the draining tank equal to ½H. If the initial value of h were different from ½H, it would take some time to attain this value.

GENERALIZING THE APPROACH TO MODEL DEVELOPMENT

We have gone step by step to a more complex and more general and powerful method of analysis. To go still further, by using a fundamental principle combined with a constitutive relationship, consider the procedure set out in Figure 3.2 for the example of the draining tank. The process described in the figure provides guidelines for proceeding when the situation is more complex than what we have considered in the context of our draining tank. As we proceed, we will consider more complex examples and illustrate the methodology and the meaning of the terms stated in Figure 3.2.

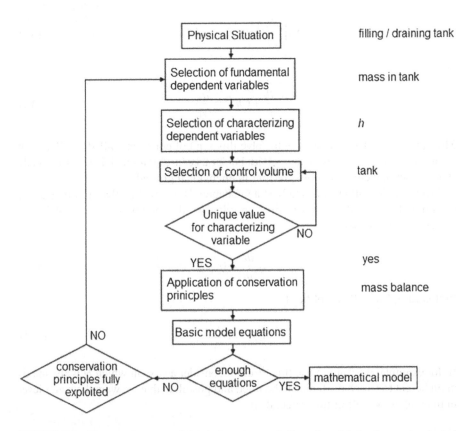

FIGURE 3.2 Systematic approach to engineering analysis and model development, adapted from a figure in T. W. F. Russell and M. M. Denn in *Introduction to Chemical Engineering Analysis* (Wiley, 1978). At the right in the figure are specifics to illustrate how the methodology was applied in this chapter to the draining tank with an inflow stream. Reproduced with permission of the publisher (Wiley).

USING MASS BALANCES

We have seen for simple examples involving single components how the mathematical statement of the law of conservation of mass provides a foundation for analysis of flowing systems. Now we generalize the approach, by including more streams and more components. Thus, for example, we might consider a tank into which three streams flow, one containing one component (say water), another containing the same component, and a third containing another component (say sodium chloride). The principle of conservation of mass must hold, and we write a balance equation with terms for the total mass flowing into the system and for the total mass flowing out.

Thus, if we had a series of tanks connected to each other and to a source of water (such as a system for purifying recycled wastewater or water from storm drains), we could write equations describing all the flows. A simple system of such tanks is shown in Figure 3.3.

But we can go further: not only is total mass conserved, the mass of water is conserved and the mass of sodium chloride is conserved, and so to take advantage of this information, we will write a mass balance equation for each of the components in addition to that for the overall mass.

To proceed, it is helpful to consider further the system (control volume) over which the balance equations will be written. Often it is more of a challenge to define this system appropriately than to write the equations.

OVERALL MASS BALANCES AND THE IMPORTANCE OF CHOICE OF CONTROL VOLUMES

A control volume is a space that may be defined in any way imaginable. The boundaries of this space define the system over which the accounting statement about mass

FIGURE 3.3 System of tanks for storing water at a water purification facility in the Salinas Valley in California. Notice the piping and valves to allow flow into the tanks at the top and out of the tanks near the base. Can you do an approximate estimate of the volume of each tank, assuming that the fence is 6 ft high? *Source*: Professor Yoram Cohen, University of California, Los Angeles, reproduced with permission.

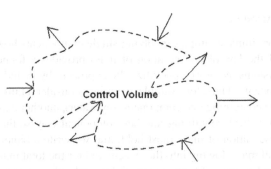

FIGURE 3.4 An arbitrary control volume over which mass balance equations can be written. There will be an equation for each component in each stream, indicated schematically by the arrows showing flow into and out of the control volume.

is written; Figure 3.4 is an example. A control volume may have any shape or size, and it may even move or change in shape or size. The choice of the control volume is sometimes obvious (as it was for our tanks), but the more complex the phenomena and the greater the number of streams, the more challenging and important it is to make an efficient definition of the control volume. The choice of this volume will determine the streams flowing in and out and thus the number, magnitude, and complexity of the terms in the mass balance equation written around the system. The system should be defined to make the mass balance equations as simple as possible, while still allowing us to obtain the information we need to model what is occurring in the system.

In a control volume, the total mass is always conserved:

$$\text{Mass Acumulation} \quad = \quad \text{Mass In} \quad - \quad \text{Mass Out} \tag{3.13}$$

This equation is now referred to as an overall mass balance, because it refers to total mass in, total mass out, and total mass accumulated. We have added the word "overall" to modify the term "mass balance," because we will proceed to consider individual components as well and write separate mass balance equations for them too—because the mass of each component is now assumed to be conserved. But we know, of course, that if a chemical reaction occurs, then this restriction no longer pertains—we consider chemical reactions in Chapter 5.

Figure 3.5 is a schematic representation of a mixed tank with an inflow stream at the top and an outflow stream at the bottom. Three different control volumes are shown, defined by the various boxes enclosed by dashed and dotted lines. We will see how important it is to choose the control volume thoughtfully.

The overall mass balance equation for the smallest control volume (A in Figure 3.5), which is just part of the inlet pipe, is as follows:

$$\frac{dM}{dt} = F_{in} - F_{out} = 0 \tag{3.14}$$

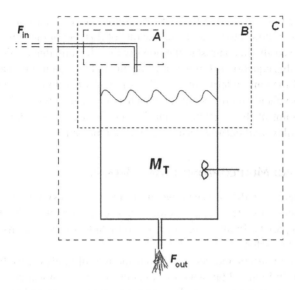

FIGURE 3.5 Schematic representation of a mixed tank (the propeller symbolizes the mixer) with an inflow and an outflow stream. Three control volumes are shown, each enclosed in dashed or dotted lines.

where M is the mass in this control volume, F_{in} is the rate of flow of mass into the control volume, and F_{out} is the rate of flow of mass out of the control volume. The mass balance equation for this control volume is very simple, but it fails to tell us anything about what is going on in the tank. If our goal is to analyze the latter, this equation is not helpful. The choice of control volume was not satisfactory.

Let us do better by considering control volume B in Figure 3.5, which is larger, but which we might anticipate will also be insufficient because it includes only part of the tank. The mass balance is as follows:

$$\frac{dM_B}{dt} = F_{in} \pm \begin{array}{c} \text{Flow associated} \\ \text{with mixing} \end{array} \qquad (3.15)$$

Here, M_B is the mass of liquid in control volume B, and we have written $\pm$ in front of the flow-by-mixing term to recognize that the mixing causes flow both into and out of the control volume. Although this balance equation includes more information than Eq. (3.14) and at least includes recognition of the mixing in the tank, it is still not useful, because it would be difficult to determine all the microscopic chaotic and variable flow rates arising from the mixing process.

To continue the process of being more inclusive of the parts of the overall system in which flow and mixing occur, we define control volume C in Figure 3.5, which now includes the whole tank. We write the following mass balance equation:

$$\frac{dM_T}{dt} = F_{in} - F_{out} \qquad (3.16)$$

where M_T is the mass of fluid in the control volume. We now have an expression that includes the rate of flow into the tank and the rate of flow out of the tank, both of which we can measure easily (in contrast to the flows within the tank that result from the mixing). Although this equation is limited, it is useful, and we can use it to account for the performance of the tank in terms of what flows in and what flows out, even though we do not account explicitly for the details of what happens in the tank (such as the mixing).

The main point of this section is that the choice of control volume is crucial; it must include inflow and outflow terms that we can measure.

MULTISTREAM AND MULTICOMPONENT MASS BALANCES

In principle, there is no difference between the analysis of a system with two streams and the analysis of a system with hundreds of streams, or more; the difference lies only in the complexity. In any case, we need to include all the streams that cross the boundaries of the system.

To complicate matters and generalize our method of analysis, we begin with the example shown in Figure 3.6b, which is a system with two inflow streams and three outflow streams (contrast this with the simpler example shown in Figure 3.6a, which we have already considered). The inflow streams in Figure 3.6b are labeled A and C, and the outflow streams are labeled B, D, and E. By extension, applying the same procedure as before, we write the overall mass balance for this system as follows:

$$\frac{dM_T}{dt} = F_A - F_B + F_C - F_D - F_E \qquad (3.17)$$

Now, we take a step to generalize this procedure by introducing a more compact notation by using symbols indicating summations. Thus, for a system with any number of inflow streams and any number of outflow streams, we write the overall mass balance as follows:

$$\frac{dM_T}{dt} = \sum_i^{\text{Inlets}} F_i - \sum_j^{\text{Outlets}} F_j \qquad (3.18)$$

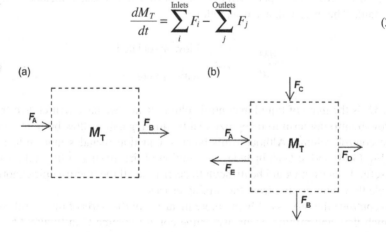

FIGURE 3.6 (a) System with one inflow stream and one outflow stream. (b) System with two inflow streams and three outflow streams. M_T is the total mass in a system. The control volumes are enclosed in dashed lines.

The summation symbols Σ represent, in the first term on the right-hand-side of the equation, an accounting for each of the inflow streams designated with the index i and, in the second term on right-hand-side of the equation designated with the index j, each of the outflow streams. For convenience, we use a convention whereby i denotes input streams and j denotes output streams.

In this representation (and the ones that follow), each arrow in Figure 3.6 represents a single stream entering or leaving the system.

It is often convenient to express the mass balance in terms of volume flow rates and densities, because these are often easier to measure than mass flow rates. Thus, each mass flow rate can be expressed as follows:

$$F_i = \rho_i \times q_i \tag{3.19}$$

where q_i is the volume flow rate of stream i and ρ_i is the density of that stream.

With this definition, the general mass balance can now be written as follows:

$$\frac{dM_T}{dt} = \sum_i^{\text{Inlets}} \rho_i \times q_i - \sum_j^{\text{Outlets}} \rho_j \times q_j \tag{3.20}$$

The next example illustrates the approach for a mixing device.

Example 3.1 Analysis of Mixing of Streams of Different Compositions

Problem statement: A stream of 50% by volume isopropanol (isopropyl alcohol) and 50% by volume water is to be prepared by mixing a stream containing only water with a stream containing 80% by volume isopropanol and 20% by volume water in a tank, with all the operations carried out at 298 K and atmospheric pressure. The tank is perfectly mixed with a mechanical stirrer, and it operates at steady state. The flow rate of the desired product is 1,000 kg/h. Consider steady-state operation. Recognize that the mixing of two streams is, roughly speaking, the opposite of separation of a stream into streams with different compositions. It is almost always much easier and less costly to mix streams than to separate streams into components.

- A. Sketch the system.
- B. Identify each of the terms in the overall mass balance for the tank. Identify each of the terms in the component mass balances for isopropanol and water. Describe each term in words.
- C. Calculate the mass flow rates of the streams entering and leaving the tank, and show how to calculate the volume flow rates.

Solution

Below is a diagram depicting the system with the given information, where i refers to isopropanol and w to water; the streams are labeled 1, 2, and 3:

Because the system operates at steady state, there is no accumulation, and the mass balances simplify to the form in $-$ out $= 0$

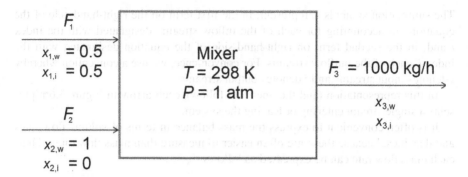

Overall balance written for the tank as the system: $F_1 + F_2 - F_3 = 0$.
Individual component mass balances:

Water: $(x_{1,w} \times F_1) + (x_{2,w} \times F_2) - (x_{3,w} \times F_3) = 0$
Isopropanol: $(x_{1,i} \times F_1) + (x_{2,i} \times F_2) - (x_{3,i} \times F_3) = 0$

where F_1, F_2, and F_3 are the total mass flow rates of Streams 1, 2, and 3, respectively; $x_{1,w}$, $x_{2,w}$, and $x_{3,w}$ are the mass fractions of water in Streams 1, 2, and 3, respectively; and $x_{1,i}$, $x_{2,i}$, and $x_{3,i}$ are the mass fractions of isopropanol in streams 1, 2, and 3, respectively.

Information provided includes the statement that outlet Stream 3 contains 50% by volume water and 50% by volume isopropanol. We can use this information to determine the mass fractions of these components in this stream. First, we need to find the densities of the two components at 298 K and 1 atm from handbooks stating these values: $\rho_{water} = 997\,kg/m^3$ and $\rho_{isopropanol} = 786\,kg/m^3$. We will assume that there is no change in total volume when isopropanol and water are mixed—that is, we assume, for example, that mixing of 1 L of isopropanol with 1 L of water gives 2 L of solution (this approximation is not perfect, but we make it here to keep matters simple). Using the density values, we find

$$F_{3w} = \rho_w \times q_{3w} = 997\ kg/m^3 \times q_{3w}$$
$$F_{3i} = \rho_i \times q_{3i} = 786\ kg/m^3 \times q_{3i}$$

A similar set of equations can be written for Stream 1, and we proceed for both of these streams to calculate the mass fractions of each of the two components; we illustrate for Stream 3, realizing that the mass fraction of a component = mass flow rate of component/total mass flow rate of stream and that (by our assumption of the lack of volume change on mixing of isopropanol and water) the volume flow rate of each of the two components in Stream 3 is the same ($q_{3w} = q_{3i}$); thus, with a few steps left out, we find that

$x_{3,w} = 49{,}850\ kg/h/(49{,}850\ kg/h + 39{,}300\ kg/h) = 49{,}850/89{,}150 = 0.56$
$x_{3,i} = 1 - x_{3,w} = 0.44$

Thus, we now know the mass fractions associated with a 50/50 by volume mixture of water and isopropanol, and therefore the mass fractions of the output stream. Doing a similar calculation for Stream 1, we find that the mass fractions of isopropanol and of water in that stream are 0.76 and 0.24, respectively. Using the balance equations stated above, we next solve for the values of the total mass flow rates of Streams 1 and 2, F_1 and F_2. For example,
    Isopropanol mass balance:

$(0.76 \times F_{1i}) + (0 \times F_{2i}) - (0.44 \times 1{,}000\ kg/h) = 0$
$F_{1i} = 580\ kg/h$

A similar calculation for the water mass balance leads to the result that $F_{1w} = 184$ kg/h.

Overall mass balance:

Using the values calculated above, we find that $F_{1i} + F_{1w} + F_{2w} = 1000$ kg/h and that $F_{2w} = 236$ kg/h.

To find the volume flow rate of each stream, we just convert the individual mass flow rates into volume flow rates by using the densities. For example,

$$q_{1,w} = F_{1w}/\rho_{water} = (184 \text{ kg/h})/997 \text{ kg/m}^3 = 0.19 \text{ m}^3/\text{h}$$

SEPARATIONS DEVICES

Let us now consider the opposite of mixing of streams and examine the separation of mixtures into components. Mass balances are used to analyze the performance of many kinds of devices and processes in which separations take place. For example, it is common in the chemical and fuel refining industries to separate mixtures into components by distillation.

A distillation column is a device that often incorporates a stack of plates with holes in them that allow vapors formed by heat added to the system to rise from one plate to the next plate above and liquid to fall to the next plate below. Both vaporization and condensation take place on each plate. The gas-liquid mixture is thus enriched in the more volatile components in the upper plates in the column and more enriched in the less volatile components in the lower plates in the column. In the simplest case, a distillation column is used to separate two components; if they are similar in volatility, the column will require many plates, but if they are widely different in volatility, it will require only a few. Some distillation columns contain dozens of plates, some are several stories tall, and some are used to separate highly complex mixtures, such as crude oil (but the separations do not give pure compounds, but rather mixtures with narrow boiling ranges). To remove the purified vapor products from a distillation column, they are usually condensed to give liquid streams.

Example 3.2 Analysis of Performance of a Distillation Column

Problem. A mixture of benzene and toluene (50% by mass of each) is partially separated in a continuous, steady-state distillation column that is fed at a rate of 1,000 kg/h. The mass flow rate of the stream flowing from the top of the column through a condenser (called the overhead stream) is 600 kg/h. The mass fraction of toluene in the bottoms stream is 0.1 kg of toluene per kg of total mass:

    Problem statement A. What is the flow rate of the bottoms stream?

    Problem statement B. What is the composition of the overhead stream?

Solution

First, we summarize the given information in a schematic flow diagram, where each term F is a mass flow rate and each term x is a mass fraction. The subscripts 1, 2, and 3 refer to the inlet, overhead, and bottoms streams, respectively, and the

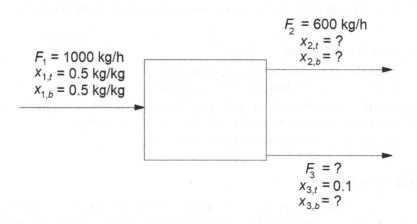

subscripts b and t refer to benzene and toluene, respectively. The question marks indicate the terms we are asked to evaluate. As usual, each arrow represents a single stream.

Consider the process in the above schematic diagram and start with an overall steady-state mass balance to find the flow rate of the bottoms stream, where m is the total mass of the system:

$$\frac{dm}{dt} = F_1 - F_2 - F_3 = 0 \tag{3.21}$$

Substituting in the given information, we find that

$$F_3 = 1000 \text{ kg/h} - 600 \text{ kg/h} = 400 \text{ kg/h} \tag{3.22}$$

Next, recognizing that each component is neither created nor destroyed (because there is no chemical reaction), we use a component steady-state mass balance to find the mass fraction of toluene in the overhead stream (we could have used the component steady-state mass balance for benzene instead of that for toluene—can you show how that leads to the equivalent result?):

$$\frac{dm_t}{dt} = x_{1,t}F_1 - x_{2,t}F_2 - x_{3,t}F_3 = 0 \tag{3.23}$$

Thus,

$$x_{2,t}(600 \text{ kg/h}) = (0.5)(1000 \text{ kg/h}) - (0.1)(400 \text{ kg/h}) \tag{3.24}$$

And so,

$$x_{2,t} = 0.77 \tag{3.25}$$

Using the fact that the sum of the component mass fractions must equal unity, we find $x_{2,b} = 0.23$ and $x_{3,b} = 0.9$.

GOING BEYOND MIXING AND SEPARATION: PROCESS FLOW DIAGRAMS

The development above illustrates the value of mass balances in analyzing processes of mixing and the reverse, separation. These are important operations in many processes, but they are not the only ones—often the processes involve chemical change as well. To begin to conceptualize processes involving chemical change, we need to recognize how operations such as pumping (flow), mixing, and separation are combined in processes in which chemical change occurs. Representations of processes (with or without chemical change) are known as process flow diagrams, and we have already made a start toward creating them. The following example illustrates a process in which chemical changes occur; the device (vessel) in which the chemical changes take place is called a reactor. In process flow diagrams, we again represent each single stream with an arrow.

In many chemical processes, tanks called surge tanks are used in the system so that constant flow rates can be maintained downstream of the surge tank even when the flow rates into the tank vary, such as might result from upsets in feed availability.

Example 3.3 Process for Methanol Synthesis

Problem statement: A process for the manufacture of methanol (methyl alcohol) (CH_3OH) from a mixture of $H_2 + CO$ (referred to as synthesis gas, or syngas) is carried out in a flow reactor, and the product is purified downstream of the reactor. The process is fed continuously at steady state with H_2 flowing at a rate of 2,000 mol/h and with CO flowing at a rate of 1,000 mol/h. The product from the reactor flows to a distillation column. Flowing from the bottom of the distillation column is a stream containing 900 mol/h of methanol, 50 mol/h of H_2, and 25 mol/h of CO. Flowing from the top of the distillation column is a stream containing H_2 and CO, and this stream is recycled to the reactor—that is, it is sent back to the reactor, mixed with the feed to the reactor.

   A. Figure out the stoichiometry of the methanol synthesis reaction.
   B. Determine whether the feed composition corresponds to the stoichiometric ratio of H_2 to CO.
   C. Determine the conversion of the synthesis gas to methanol—that is, determine the fraction of the H_2 (and of the CO) that is converted to methanol in the steady state.
   D. Draw a simple flow diagram to represent the system, showing each of the streams and labeling each stream with the components flowing in it.

Solution

A. and B. The methanol synthesis reaction has the stoichiometry $CO + 2H_2 \rightarrow CH_3OH$, and, therefore, a stoichiometric (molar) ratio of CO to H_2 fed to the reactor is 1 to 2, which corresponds to the given ratio of 1,000/2,000.

   C. To determine the conversion of the reactants, we need to analyze what flows into the system and what flows out. We represent the system with the flow diagram shown below, consisting of a mixer, reactor, and separator. The streams include the feed and product. Another stream (Stream 4) consists of components including CO and H_2 that are not converted in the reactor and returns them to the reactor for a further opportunity to convert them. This is called a recycle stream.

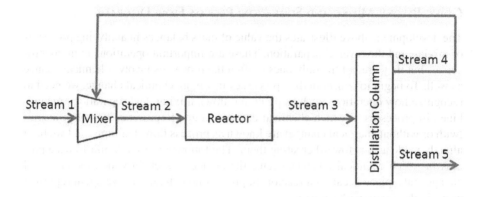

We are given the molar flow rates in Streams 1 and 5. We see that the flow rate of CO into the system is 1,000 mol/h and the flow rate of CO out of the system is 25 mol/h. And so the fraction of the CO converted is $(1,000 - 25)/1000 = 0.975$. The fraction of the H_2 converted is $(2,000 - 50)/2,000 = 0.975$. We emphasize that the calculations here involve the reaction stoichiometry and not mass balances. It is left as an exercise for the student to calculate the flow rates and compositions of Streams 2, 3, and 4.

The representation of process flow diagrams can be extended beyond those used in chemical processing. For example, let us extend the ideas to the steps in processing in an example that has been known since ancient times, when small batches were typically used and specialized equipment was assembled from natural resources. An example involves the production of food from acorns as practiced by the Native Peoples of California and neighboring areas. Acorns were an excellent food resource because of their wide availability from a number of species of oak trees, storability, and nutritional value, including substantial proteins, carbohydrates, fats, and essential vitamins and amino acids. It has been estimated that well more than half of the original Californians relied on acorns as a daily food source. An acorn fragment is shown in Figure 3.7.

These people constructed granaries to store acorns for as many as about 10 years, accumulating an excess in good years and surviving the lean years. These granaries provided what we refer to as a surge capacity. They used specially designed baskets to sort and store acorns at various stages of processing. They ground the acorns into fine powder using mortars that were holes in granite rocks (Figure 3.8) and rock pestles that, with the holes, were shaped by years of grinding. One can find many such grinding rocks in parks, such as Yosemite National Park; they were usually near streams or springs. The early people wove baskets to sieve the powder and separate the fine-grained meal from other components. Then another step was needed: because acorn meal has a high content of bitter tannic acid, it was treated in a separation step: leaching, with hot or cold water to remove the colored tannic acid. The hot-water leaching was done in baskets, with the solution heated by placement of hot rocks in the baskets. The tannic acid could be partially removed in a first leaching step, and then, with fresh water, a

FIGURE 3.7   Acorn fragment from an oak tree near the grinding rock shown in Figure 3.8.

FIGURE 3.8   Native American grinding rock. The rock is granite and located near a water source in the Sierra Nevada mountains. The tree in the background is an oak. The length of the pestle is about 21 cm; it has a flat base (do you understand why?). Smaller pestles were used with the smaller mortars.

second step was carried out and repeated until the solution was almost colorless, indicating removal of almost all the tannic acid. The resulting product was eaten as soup, mush, or heated on hot rocks to make bread. There was an important social component to these activities. Beginning with the fall harvest, women and men worked together to knock acorns from trees with sticks or simply pick them up from the ground. Women gathered the acorns into large baskets and carried them to encampments or villages where the processing was done.

Example 3.4 Process for Converting Acorns into Bread

Problem statement: Identify the steps in the process for converting acorns into bread that involve the following: flow, mixing, separation, heat transfer, and chemical change.

Solution

Flow: Water was poured from baskets into the baskets containing the ground acorn meal, and the resulting tannin-containing solutions were poured out of the latter baskets.

Mixing: The action of the mortar and pestle mixed the solid particles formed from the acorns (as well as reducing the particle sizes). When the leaching was occurring, the solid and liquid were mixed, and the mixing increased the rate of extraction of the tannic acid. The early California people made tools for the mixing (can you find information about them?).

Separation: The fine flour was separated from larger pieces of acorn fragments by sifting through fine-mesh baskets. The water–tannic acid solutions formed in leaching were removed from the solids by pouring off the solution (decanting).

Heat transfer: Substantial heating of the solutions resulted from addition of rocks to the baskets containing the solutions; heat was transferred to the rocks from fire, and heat was transferred to the mixture of solution and meal in the basket to make the leaching proceed faster than it would have when cold water was used (sometimes it was). Heat was transferred from hot rocks to the baking bread.

Chemical change occurred when the hot-water extraction was carried out, and also when the bread was baked.

Can you sketch a process flow diagram for this sequence of steps? Can you do some research and find out how to obtain acorn flour to make your own bread? If you are interested in seeing photographs of people carrying out these steps in the traditional ways, seek out online information from the Yosemite Research Library of Yosemite National Park.

Example 3.5 Process for Manufacturing Aldehydes from Olefin, Hydrogen, and Carbon Monoxide

Problem statement: Examine the process flow diagram shown in Figure 3.9 for industrial manufacture of aldehyde from olefin, hydrogen, and carbon monoxide. Figure out what happens in the pieces of equipment labeled reactor, distillation column, phase separator, distillation column, and heat exchanger.

Solution

The symbols representing the reactor include a stirrer motor (a circle) and a mixer, evidently driven by the stirrer motor. The fluid in the reactor is well mixed. The feed streams into the reactor include one that under ambient conditions (atmospheric pressure and a temperature near room temperature) is gas, consisting of carbon monoxide and hydrogen. Another stream is liquid water. And so we see that the reactor contains gas and liquid phases. The mixer serves to maintain small bubbles in the reactor to facilitate dissolving of the carbon monoxide and hydrogen in the liquid. We recognize that one of the reactants, the olefin, and the product of the reaction, an aldehyde, are organic compounds and recognize that they may not have high solubilities in water. The stirrer helps to keep one liquid phase

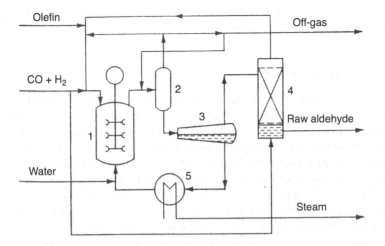

FIGURE 3.9 Process flow diagram for manufacture of aldehyde from olefin, hydrogen, and carbon monoxide. Identification of symbols: (1) reactor, (2) separator (distillation column), (3) phase separator (settler), (4) distillation column (separator), and (5) heat exchanger. Reproduced with permission from H. Bach, W. Gick, W. Konkol, and E. Wiebus, *Proc. 9th Int. Congr. Catal.* (Calgary), **1988**, Vol. 1, p. 254.

dispersed as drops in the other liquid phase and to provide a large surface area between the two liquid phases. This design allows rapid transport of reactants between the separate liquid phases and allows transport of the reactants to the phase that contains the other reactants.

The distillation column (separator) marked with the number "2" separates components by boiling point. The highly volatile carbon monoxide and hydrogen go overhead. These unconverted reactants in the stream called the overhead stream are recycled to the reactor. The products and unconverted reactants that flow downward from the separator (called the bottoms stream) flow to another separations device (3), as is evident because two streams flow from it. Separator 2 is a distillation apparatus. It may be very simple and is sometimes called a flash pot, just separating components in streams that have widely differing boiling points.

The symbolism of the separator marked "3" indicates that it separates two liquid phases that are barely soluble in each other. This is sometimes called a settler, because the denser liquid phase settles to the bottom and the less dense one rises to the top. Over time, the drops disappear as the liquids separate into their separate phases. What happens in this device is what happens in a separatory funnel used in a laboratory. We see that the heavier phase contains water and is recycled to the reactor. Some low concentrations of unconverted reactants and products would be present in this stream. The less dense liquid phase contains the aldehyde and some water—to be purified by distillation downstream in the device labeled "4." We see that some water must be present in this stream, which explains why some water ("makeup water") is introduced into the reactor to compensate. The reaction takes place in the presence of a compound that makes the reaction go faster than it would otherwise—it is called a catalyst. The catalyst is recycled to the reactor.

Can you understand why it would be advantageous to have two settlers operating in the system, so that the two liquid phases have time to separate in one of them while the other is being filled? We refer to such an operation alternating the operation of separate devices as a "swing" operation. The settlers would provide surge capacity in the flow system.

Recap and Review Questions

We have now created a foundation on which to start building an analysis in a general way, by stating the general principle known as the conservation of mass, which takes the form of a balance equation. When combined with experimental results in the form of constitutive relationships, the mass balance equation provides a framework for proceeding with model development. The procedures are further generalized in Figure 3.2. Can you trace our development of the tank-draining model through each step of Figure 3.2? Suppose you had a system consisting of the draining of one tank into another; how would you set out to analyze it by using mass balances? Beyond this, we have considered processes in which flow, mixing, and separation take place and learned how to represent them in process flow diagrams, which give insight into where various streams flow and how the devices in which mixing, separation, and chemical reaction take place work together.

PROBLEMS

3.1. A tank like the one considered in Chapter 2 that drains through an orifice in its base was used to generate data for h vs. t as the tank drained (h is the height of water above the orifice, and t is the time when the draining started). The tank is cylindrical in cross section, and the cross-sectional area A is 1,662 cm². Two identical tanks (called the upper tanks) are mounted above this tank (now called the lower tank), so that all the water that drains from the upper tanks flows into the lower tank. There is also a water pipe with a valve that can be used to feed water into the lower tank at a constant, but adjustable, volume flow rate. The valve allows adjustment of the flow rate of water into the lower tank.

A. Sketch the flow diagram and identify each of the terms in the mass balance for each of the upper tanks; describe each term in words.

B. Write the mass balance equation for the lower tank, using the following symbols and assuming that at time $t=0$ one of the upper tanks is full, the other upper tank is half full, and the lower tank is half full (and that the flow rate of water from the pipe is low enough that the lower tank never overflows—if this is possible):

q_f = volume flow rate of feed water from pipe; assume this is a constant.

q_{ut} = volume flow rate from the upper tank.

q_{out} = volume flow rate out of the lower tank.

ρ = density of water.

C. Identify each term in each equation in words and state which terms vary with time.

D. Write the overall mass balance equation for each tank by including our experimental result showing that the rate of flow of water from the tank is proportional to the square root of the height of liquid above the orifice. Use this relationship for each tank, but note that your final mass balances should be dependent only on the height of the liquid in the bottom tank, time, and the initial heights of both tanks.

3.2. Two streams with control valves provide feeds into a tank. The tank originally contains 100 kg of water. Stream A contains 0.60 mass fraction sodium sulfate, with the remainder being water. Stream B contains 0.40 mass fraction sodium carbonate, with the remainder being water. When the valves are open, feed stream A is supplied at a rate of 40 kg/h, and feed stream B is supplied at a rate of 60 kg/h. (Atomic weights are as follows: H:1, O:16, S:32, C:12, Na:23; units: g/mol).

A. Stream A was fed into the tank (initially containing 100 kg of water) for 0.5 h, and then Streams A and B were fed into the tank simultaneously for another 1.5 h. The tank did not overflow. List all the components that were present in the tank at times 0.5 and 2 h. Establish the mass balance equation for sodium sulfate in the system (where the system is the tank). At 1.5 h, what is the mass fraction of sodium in the tank?

B. After 2 h, the drain pipe from the tank is opened to allow an outflow at a flow rate $\dot{m}$, where $\dot{m}$ has dimensions of mass/time. To reach a steady state in the tank, what is the total mass flow rate of sodium ions out of the tank?

3.3. A circular cone is installed so that its tip is next to the floor and its axis is vertical. The radius at the top is $R=2$ m, and its height is 7 m. The cone is filled to the top with water. The water can be pumped through a tube out of the cone; the tube is connected to the cone near the tip of the cone. However, there is an automatic shutoff that stops the operation when the height of the liquid is 2 m above the tip of the cone. The flow rate out of the tank is a constant, $0.1\,m^3/min$.

A. How long will this run before the operation is stopped by the automatic shutoff?

B. Derive an equation (model) for the liquid height as a function of time assuming that the outflow rate is $0.1\,m^3/min$. (Note: The volume of a cone is $V=(1/3)\pi R^2 h$, where R is the radius of its base and h is its height; be careful when you do the second part—note that R depends on h.)

3.4. A scientist wants to test a growing plant's sensitivity to carbon dioxide (CO_2) by exposing it to a gas mixture of nitrogen (N_2), CO_2, and water vapor (H_2O). To make up this gas stream, the scientist has access to a pure N_2 feed and a pre-mixed CO_2/H_2O feed stream that includes CO_2 flowing at a rate of $1.17\,kg/m^3$ and an H_2O stream flowing at a rate of $0.13\,kg/m^3$. The gas stream to be sent to the growing chamber is needed at a rate of 1.0 kg/h, and it should contain the same mass fraction of CO_2 as is found in the earth's atmosphere.

A. What is the composition of the gas to be fed to the growing chamber?

B. What is the mass flow rate of the CO_2/H_2O feed?

C. What is the composition of the final mixed gas stream?

D. What is the mass flow rate of the N_2 feed stream?

3.5. In a continuous steady-state mixing process, a concentrated isopropanol solution is diluted with pure water. The isopropanol feed stream is 80% isopropanol and 20% water (by mass), and it is supplied at a rate of 62.5 kg/h. The pure water stream is supplied at a rate of 37.5 kg/h. What is the total mass flow rate of the output stream, and what is its composition (by mass)?

3.6. Prepare a process flow diagram for converting coffee beans into coffee. Identify steps in which grinding, flow, mixing, extraction, heat transfer, and chemical change occur. Compare the process of coffee making with the process of making soup from acorns. What can you learn about the history of coffee making? What social issues can you identify related to harvesting of coffee beans?

3.7. A tree called a piñon pine (sometimes written "pinyon pine") in the Southwest United States was an important food source for Native Americans. Do some research and find out how the pine nuts were harvested and processed and why they were such a valuable food source. Sketch a process flow diagram that includes the harvesting and processing steps. Find out what other products of piñon pines were valuable to the Ancient People. How did one of these products help in the transport of water? What can you learn about how the work of harvesting piñon pine nuts was distributed between men and women?

3.8. Methane is a greenhouse gas that is having a major impact on the earth's climate. Identify the major (natural and human-induced) sources of methane that enters the earth's atmosphere and the major sinks, where methane is removed from the atmosphere. Go to the literature and find estimates for an approximate mass balance calculation that indicates the approximate current rate of increase in the atmospheric methane content. Also estimate the methane concentration in mol/L in the earth's atmosphere at sea level.

3.9. Go online and find out more about the water purification facility in the Salinas Valley depicted in part in Figure 3.3.

 A. Find a process flow diagram for the facility and explain it; explain why the pumps and valves are placed where they are. Explain what surge capacity means and which tanks provide it.

 B. What community is the facility designed to serve? Read about the facility and summarize important points about its social significance and relationship to public health.

3.10. Find out what polychlorinated biphenyls (PCBs) are.

 A. Do some research and find out why they are a public health concern.

 B. Check some literature starting with the article by M. S. McLachlan in *J. Agric. Food Chem.*, **1993**, *42*, 474–480. Show a mass balance for a lactating cow, with quantitative information accounting for PCBs in milk that is consumed by humans. Do not overlook an accumulation term.

3.11. A significant source of the greenhouse gas methane is cows. Do some research and present an approximate mass balance for an adult cow (a female), averaged over a long time. Account for the input food and oxygen from the air and account for the outputs that are gases, liquids, and solids. Do not overlook the accumulation term.

3.12. Find out what gluten is and devise a process for making gluten from wheat. Present a process flow diagram for the process. Explain why gluten-free products are important. Can you find any mass balance data characterizing the process?

3.13. Find out what a resurrection plant is. Get one (they are available online) and use a camera and a stopwatch to characterize its growth. Suggest measures of its size, and plot them as a function of time. Suggest an equation to represent the rate of growth.

3.14. Find out what are the major sources of synthesis gas and do the research about how synthesis gas is made from coal and how that process contributes to global warming. Can you find any mass balance data characterizing the process, including data for products that are different from synthesis gas (such as inorganic products sometimes called fly ash)?

3.15. Fish caught from streams at relatively low elevations in the United States contain contaminants that are hazardous to human health. Do some research and identify some such contaminants and explain how they are assimilated by fish. Can you find any mass balance data characterizing the contaminants?

4 Units, Dimensions, and Dimensional Analysis

ROADMAP

The dimensions that are characteristic of all physical quantities provide insight into physical processes. We review the basic ideas of dimensions and units, showing the usefulness of the fact that each term in an equation must have the same dimensions. The essential point of this chapter is that, solely through analysis of the dimensions, it is possible to predict the forms of equations representing physical phenomena; the process is called dimensional analysis. Terms that are combined to form dimensionless groups appear frequently in the analysis of physical phenomena and sometimes provide simple criteria for such phenomena, illustrated by laminar and turbulent flow predicted by the dimensionless group called the Reynolds number. Dimensional analysis is illustrated for the draining of water from a tank through an orifice in its base, confirming the results developed in Chapter 2.

DIMENSIONS

We associate the word "dimensions" with figures and objects; examples of dimensions are lengths, depths, and radii. In geometry, two points are separated by a distance—the length of the straight line connecting them—this is a dimension. Dimensions have units; we commonly measure lengths in meters (m) or feet (ft). Periods of time have dimensions that we usually express in the units of seconds (s), minutes (min), or hours (h). Fundamental physical quantities such as the velocity of light and the velocity of sound in air have dimensions measured in units such as meters per second ($m \times s^{-1}$ [m s^{-1}]). All physical properties of gases, liquids, and solids, such as density, thermal conductivity, and viscosity, have dimensions. Density is often measured in units of kg m^{-3}.

Fundamental dimensions include length (L), time (T), mass (M), and force (F). Other fundamental dimensions, such as those encountered in electricity and magnetism, are left out of our discussion because we do not consider such phenomena in this book.

Dimensions and units are not the same; each dimension has units, but units are not dimensions, although we can figure out what dimensions pertain to a particular set of units. Various sets of units are used to represent the fundamental dimensions. The two most important sets of units for engineers are summarized in Table 4.1. These are the SI (French Système International d'Unités) system and the engineering system. Another common set is the CGS (centimeter-gram-second) system (Table 4.1), and it is commonly used by scientists. We often need to convert quantities from one system of units to another.

DOI: 10.1201/9781003429944-4

TABLE 4.1
Systems of Units

Dimension	Units in SI System	Units in Engineering System	Units in CGS System
Length (L)	meter (m)	foot (ft)	centimeter (cm)
Time (T)	second (s)	second (s)	second (s)
Mass (M)	kilogram (kg)	pound mass (lb_m)	gram (g)
Force (F)	Newton (kg m s^{-2})	pound force (lb_f)	dyne (g cm s^{-2})

We consider two types of dimensions, primary and secondary. As shown in Table 4.1, there are three *primary dimensions* in the SI system, *mass*, *length*, and *time*; all others are secondary. The dimensions of force are a combination of these primary dimensions (and they are evident directly from Newton's law of motion—do you recall what this is? Can you check this statement?). In the engineering system of units, force is considered to be an independent dimension.

Engineers primarily use the SI system of units, which in recent decades has gained ascendancy over the engineering system and the CGS system but has not displaced them.

To denote the dimensions of a quantity, we use the symbol [=], which means "has dimensions of." Thus, for example, we write the following for velocity v and acceleration a:

$$v \ [=] \ LT^{-1}$$

$$a \ [=] \ LT^{-2}$$

The former statement means "the velocity has dimensions of length per unit time" or "v has dimensions of LT^{-1}."

Some non-engineering subjects that are important to engineers require their own primary dimensions. For example, in economics, the primary dimensions are *money* and *time*. The dimensions of money must meet several mathematical tests about how it is measured and the operations it may undergo (these operations include addition and subtraction and currency exchanges—which are changes of units, say from euros to dollars). Depending on the branch of economics, additional primary dimensions may be added. For simple macroeconomics, the dimension of *goods* is included. This is a common measure against which objects or commodities are traded, such as wheat, pork bellies, and automobiles. From these primary dimensions, one may extract secondary dimensions, such as the flow of money, a measure of how fast money is transferred.

DIMENSIONAL HOMOGENEITY (DIMENSIONAL CONSISTENCY)

Here is an important and at first surprising principle: some *physical insights result simply from the analysis of dimensions*. Thus, we use what we call *dimensional analysis*, which is a simple, powerful tool to develop functional relationships among physical variables. The meaning of these words will become clearer below.

A central idea is that when variables are related through equations, *each term in the equation must have the same dimensions*; otherwise, the equation is internally inconsistent and incorrect. This is the requirement of dimensional homogeneity.

For example, consider a moving object. From physics, we know the formula for the position of an object undergoing a constant acceleration; it is derived in calculus in terms of velocity v and acceleration a:

$$y - y_0 = \frac{1}{2} a t^2 + v_0 t \qquad (4.1)$$

Here, y_0 is the initial position of the object (think of it as the distance from the origin on a graph); v_0 is the initial velocity; t is the time; and y is the instantaneous position, which is a function of time. For this equation to describe real physical phenomena, each of the four terms in the equation (y, y_0, $\frac{1}{2}at^2$, and v_0t) must have the same dimensions.

Consider this equation further. The left-hand side represents the displacement of the object from its initial position (the distance it has moved in time t). Suppose that the object is a car you are driving on a highway, and suppose you decide to pass another car traveling at a constant velocity. To pass it, you must accelerate. The displacement given by Eq. (4.1), ($y - y_0$), is the distance you would travel during the period of acceleration relative to the position from which you started the acceleration (y_0). This distance has dimensions of length (L). On the right-hand side of the equation, the initial velocity v_0 has dimensions of LT^{-1} (this is the velocity that you had attained at the instant you decided to pass (and accelerate)). When we multiply v_0 by time, we get the term v_0t, which also has the dimension of length—as it must. Furthermore, the acceleration a has dimensions of $LT^{-?}$, and multiplying this by t^2, which has the dimension of T^2, also gives the dimension of length, L—as it must. Therefore, our check shows that each term in Eq. (4.1) has the dimension of L, and the equation meets the criterion of being *dimensionally homogeneous* or *dimensionally consistent*. If it did not, it would be wrong.

Let us check the dimensions of each term in the equation to confirm the consistency:

$$y - y_0 \, [=] \, L$$

$$\frac{1}{2} a t^2 \, [=] \, \frac{L}{T^{-2}} \text{ multiplied by } T^2, \text{ or } L$$

$$\text{and } v_0 t \, [=] \, L/T \text{ multiplied by T, or L.}$$

Now let us relate what we have learned about dimensions to the draining tank and the mass balance equation that describes it. Consider, as in Chapter 2, a tank of cross-sectional area A, in a gravitational field represented by g:

$$\frac{d}{dt}(\rho A h) = -\rho q \qquad (4.2)$$

Here, as before, the symbol q represents the volume flow rate. The fluid density is represented by ρ [=] ML^{-3} and is a constant in the draining tank; h [=] L is the height of the liquid above the orifice at the base of the tank; and q [=] L^3T^{-1}. Let us check this equation for dimensional homogeneity: in Eq. (4.2), realizing that the mathematical operator d/dt has dimensions of T^{-1}, we see that the corresponding dimensions of the individual terms are $T^{-1}ML^{-3}L^2L$ on the left-hand side and $ML^{-3}L^3\,T^{-1}$ on the right-hand side.

The dimensions on both sides of the equation thus reduce to MT^{-1}, and, therefore, the equation is dimensionally homogeneous. The concept of dimensional homogeneity is central to all physical theories or models. It is useful as a check of our manipulations of equations; if we use algebra to rearrange or combine equations, then we can use the criterion of dimensional homogeneity to check our work; if the resulting equation lacks dimensional homogeneity, we have made a mistake.

We illustrate checking of dimensions further in Example 4.1.

Example 4.1   Units in an Equation of State

Problem statement: Check for the consistency of units in the SI system and figure out units of terms in the ideal gas law and in a more complex equation of state, the Benedict-Webb-Rubin (BWR) equation of state. Show that the equation of state we are familiar with, the ideal gas law, is a limiting case of the BWR equation of state.

Solution

Start with the ideal gas law, $PV = NRT$, where P is the pressure, V is the volume, N is the number of mols, T is the absolute temperature, and R is the universal gas constant. Let us write this as $P = \rho RT$, where ρ is the molar density and N/V is the number of mols/volume, to facilitate comparison with the BWR equation, which is stated below. The gas constant R has the value of 8.3144 kg m^2 s^{-2}. P has SI units of Newton m^{-2} or kg m s^{-2}m^{-2} or kg m^{-1}s^{-2}, ρ has SI units of (number of mols)(m^{-3}), and R has SI units of kg m^{2}s^{-2}.

Thus, the right-hand side of the equation, ρRT, has SI units of (number of mols)(m^{-3})$\times$(kg m^{2}s^{-2})$\times$(number of mols)$^{-1}\times K^{-1}\times K$, or kg m^{-1}s^{-2}, providing the expected confirmation by matching the units of pressure. The BWR equation is the following (where the meaning of A is different from what we used before):

$$P = \rho RT + \left(B_0RT - A_0 - C_0T^{-2}\right)\rho^2 + (bRT - a)\rho^3$$

$$+ \ \alpha a\rho^6 + c\rho^3T^{-2}\left(1 + \gamma\rho^2\right)\exp\left(-\gamma\rho^2\right) \tag{4.3}$$

First, we see that if the term ρRT is large by comparison with the sum of the other terms on the right-hand side of Eq. (4.3), then we have the ideal gas law as a limiting case. We also see that each term on the right-hand side of this equation must have the units of P, that is, kg m^{-1}s^{-2}. Now look at the exponential term at the right: the exponent of e (i.e., $-\gamma\rho^2$) must be dimensionless, and so γ must have the units of ρ^{-2}, or (m^{-3} per number of mols)^2, or m^{-6}, because the number of mols is a number, dimensionless. We confirm this conclusion by seeing that in the equation $\gamma\rho^2$ is added to a pure number, 1, in the right-hand term. Working from right to left in the equation, we see that $c\rho^3 T^{-2}$ must have the SI units of P, kg m^{-1}s^{-2}, and so c has the SI units of kg m^{8}s^{-2} K^{-2}. Similarly, $a\rho^3$ must have the SI units of P, kg m^{-1}s^{-2}, and so a must have the SI units of kg m^{8}s^{-2}.

The remaining assignments are left as an exercise for the reader.

DIMENSIONLESS VARIABLES

We now take an important step forward by introducing the idea of dimensionless variables—variables that have no dimensions (or units), such as, for example, the length to diameter ratio of a pipe or the value of π (what is the meaning of π in geometry?). Can you think of a dimensionless variable that we used in the analysis of tank draining in Chapter 2? (It appeared in a graph.)

A dimensionless variable is a (pure) number. Two of the many examples of dimensionless variables that are important in chemical engineering are the following:

The Reynolds number (characteristic of a flowing fluid):

$$N_{Re} = \frac{(\text{density})(\text{velocity})(\text{length})}{(\text{viscosity})} = \frac{\text{inertial forces}}{\text{viscous forces}} \tag{4.4}$$

The Prandtl number (characteristic of the physical properties of a fluid, heat capacity of a fluid, viscosity, and thermal conductivity, which we will introduce in Chapter 6):

$$N_{Pr} = \frac{(\text{heat capacity})(\text{viscosity})}{(\text{thermal conductivity})} = \frac{\text{viscous heating}}{\text{conduction}} \tag{4.5}$$

The right-hand side of each of these equations gives an imprecise statement of what the numerator and denominator represent physically; the meaning of the terms representing the Reynolds number is clarified below. These and other dimensionless groups are named after famous scientists and engineers who have contributed to the fields in which the numbers are central.

The dimensionless groups facilitate analysis of physical phenomena because they group terms that belong together—*in general*—as indicated by analysis of the physical phenomena. We begin to illustrate the meaning of this statement by considering an example: the flow of water through a pipe with a circular cross section. In such a situation, the velocity in Eq. (4.4) is the average velocity of the fluid in the pipe, and the length term is the pipe diameter (not the pipe length). The density and viscosity are those of the fluid in the pipe.

It has been found by experiment that in a particular pipe with a circular cross section, when the velocity corresponds to a Reynolds number of 10 (calculated according to Eq. (4.4)), the flow is *laminar*. When a drop of colored dye is injected into a fluid in laminar flow in a glass pipe, it is observed to move with the liquid in a nearly straight line, as sketched in Figure 4.1. It has been found by experiment that, for other fluids in other pipes, laminar flow is always observed when the Reynolds number is,

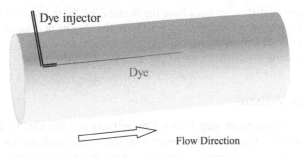

FIGURE 4.1   Laminar flow. The continuously injected dye moves with the flow (from left to right) and a clear streak line can be observed.

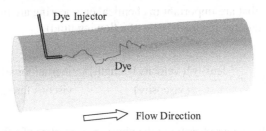

FIGURE 4.2   Sketch showing some characteristics of turbulent flow. The continuously introduced dye does not flow in a straight path. It becomes increasingly diluted as it moves down the pipe and mixes with the fluid.v

for example, 10. Other values of the Reynolds number also correspond to laminar flow. The result is general: this value of the Reynolds number tells us that the flow is laminar for all fluids (e.g., air, water, gasoline, honey) and for pipes with circular cross sections of any diameter and made of any material.

Laminar flow can occur in any stream, not just in a pipe. Have you ever observed a creek with laminar flow? How could you tell it was laminar? Did nature provide something playing the role of the dye in the example stated above?

In contrast, if the velocity in the pipe is increased so that the Reynolds number becomes 4000, for example, observations show that the injected dye is not so easy to track. It moves along the pipe chaotically—up and down and side to side as well as downstream. In other words, the flow is not just parallel to the axis (length) of the pipe, but also occurs in the perpendicular (radial) direction. Such a chaotic flow is called *turbulent* (Figure 4.2). It has also been found by experiment that turbulent flow is always observed when the Reynolds number is 4000 (among other values). Thus, this result also is general: the flow is turbulent for all fluids and all pipes with circular cross sections when the Reynolds number is 4,000.

Have you ever observed a river with turbulent flow? How could you tell that it was turbulent?

There is a value of the Reynolds number that corresponds to the transition from laminar to turbulent flow; it is about 2,100. Thus, the Reynolds number, by itself, determines whether the flow in the pipe is laminar or turbulent. Correspondingly, we

TABLE 4.2

Properties of Air and Water

Fluid	Density	Viscosity
Air	1.33 kg m^{-3}	1.73 × 10^{-5} kg m^{-1} s^{-1}
Water	62.2 lb$_m$ ft^{-3}	2.17 lb$_m$ ft^{-1} h^{-1}

say that two flows are *dynamically similar* to each other if their Reynolds numbers are the same.

In many applications, it is important whether the flow is laminar or turbulent. For example, turbulent flow causes rapid mixing of the dye in the water—so that not far downstream from where the dye is injected, it might be difficult to even see the dye, because it would be spread out (dispersed) in the fluid across the whole cross section of the pipe, and it might become so dilute as to be essentially invisible. If we wish to mix two fluids, we could do it much more effectively in turbulent flow than in laminar flow.

We now do a comparison of the Reynolds numbers for two different fluids, air and water, flowing through two different pipes. The comparison requires physical properties of air and water, as shown in Table 4.2.

The physical properties of air and water in Table 4.2 are given in different units. To compare them with each other, we need them in the same units. That is, we need to convert from one set of units to another, and, to do this, we need *conversion factors*. One of these is shown in Eq. (4.6) in the numerator on the right-hand side, where we convert inches to feet. We do this by multiplying the length in inches by the conversion factor of 1 ft/12 in. Any conversion factor is a dimensionless number. In this case, because 1 ft = 12 in, the term (1 ft/12 in) is a dimensionless number—and it has a value of exactly unity—so that when we multiply the value of length by this number, we do not change its value—all we change is its units. In the following equations, we use several conversion factors (can you identify them and check whether they are correct?).

Let us consider what ratio of velocity of air to velocity of water gives the same Reynolds number for flow in a pipe, presuming the diameter of the water pipe is 1 inch and that of the air pipe is 1 foot. Using the physical properties of air and water in Table 4.2, we find for water

$$(N_{Re})_{\text{water}} = \frac{(62.2)(v_{\text{water}})(1)}{(2.17)}[=]\frac{\left(\dfrac{\text{lb}_m}{\text{ft}^3}\right)\left(\dfrac{\text{ft}}{\text{s}}\right)(\text{in})\left(\dfrac{1\ \text{ft}}{12\ \text{in}}\right)}{\left(\dfrac{\text{lb}_m}{\text{ft h}}\right)\left(\dfrac{1\ \text{h}}{3600\ \text{s}}\right)} \tag{4.6}$$

For air, we find

$$(N_{Re})_{\text{air}} = \frac{(1.33)(v_{\text{air}})(12)}{(1.73 \times 10^{-5})}[=]\frac{\left(\dfrac{\text{kg}}{\text{m}^3}\right)\left(\dfrac{\text{ft}}{\text{s}}\right)(\text{in})\left(\dfrac{0.0254\ \text{m}}{\text{in}}\right)\left(\dfrac{1\ \text{m}}{3.28\ \text{ft}}\right)}{\left(\dfrac{\text{kg}}{\text{m s}}\right)} \tag{4.7}$$

Setting these two Reynolds numbers equal to each other, we find $v_{water} = 1.2\ v_{air}$. If this velocity ratio is maintained at 1.2, the flow regimes will be the same.

IDENTIFYING DIMENSIONLESS GROUPS

There is a formal procedure involving a number of operations that lead to identification of dimensionless groups for any physical process. We present an abbreviated form of this procedure, emphasizing three parts that are critical:

1. identifying the relevant variables that describe a physical phenomenon;
2. applying a theorem called the Buckingham Pi theorem (or the Pi theorem), which is introduced below; and
3. identifying the independent dimensionless groups.

The Buckingham Pi theorem is used to determine how many independent dimensionless groups are associated with any physical phenomenon. It underlies the powerful method of dimensional analysis, which we now introduce with an example.

Example 4.2 Motion of a Pendulum

Problem statement: Use dimensional analysis to analyze the motion of a simple pendulum, shown in Figure 4.3, and determine its period. The pendulum has length *l* and mass *m*, with all the mass assumed to be concentrated at a point. The pendulum thus moves as a point, under the influence of gravity. At any instant, the pendulum is displaced by an angle θ from the vertical, and the maximum displacement angle is α.

Solution

We use dimensional analysis to estimate the period (and hence the frequency) of the swing as the pendulum traverses from one extreme position to the other. We simplify the analysis by ignoring some forces that work on the pendulum, because they are small in comparison with the force of gravity. Thus, we assume that the pendulum experiences no forces that would retard its motion, such as the

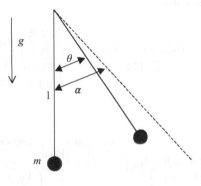

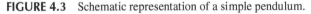

FIGURE 4.3   Schematic representation of a simple pendulum.

TABLE 4.3

Variables for Pendulum Analysis

Name of Variable	Symbol for Variable	Dimensions
Period of oscillation	τ	T
Length of pendulum	l	L
Mass of pendulum	m	M
Acceleration of gravity	g	LT^{-2}
Amplitude of oscillation	α	dimensionless

resistance offered by the air that acts on both the mass and the string to which the mass is attached (this is called viscous drag). Furthermore, we ignore other forces, such as the Coriolis force, which pushes the pendulum outside of a planar oscillation and results in the well-known Foucault pendulum.

The variables are summarized in Table 4.3. These variables and their dimensions are familiar, except for α. This is the oscillation amplitude, measured in radians; it is the ratio of the arc length to the radius and is therefore dimensionless. On the basis of physical reasoning alone, we anticipate that the period of oscillation τ should depend on the other variables. We do not know in advance what the dependence is. Finding the precise relationship is a task for a physicist. Here, we simply seek to apply our basic analysis skills to find out how these variables depend on each other. We represent the dependence through an arbitrary and—at this point—still unknown function, f, as follows:

$$\tau = f(l, m, g, \alpha) \tag{4.8}$$

This equation states only that the period of oscillation depends (somehow) on the length and mass of the pendulum, the force of gravity, and the amplitude of the motion of the pendulum. The term on the left-hand side of this equation has dimensions of time. The several variables on the right-hand side have dimensions of time, length, and mass. The result of operating on these variables by the function f must be to create a grouping of the variables so that the outcome has dimensions of time.

To find out how the variables can be arranged to give a grouping with the dimensions of time, a good way to start is by noticing that we can combine l and g, as (l/g), to obtain a variable that has the dimensions of T^2 (without L), as shown by the following:

$$\left(\frac{l}{g}\right) [=] \left(\frac{L}{\dfrac{L}{T^2}}\right) [=] T^2 \tag{4.9}$$

Now consider the mass; none of the variables other than m itself has dimensions that include mass. Therefore, it is not possible to find another variable to combine with m to eliminate m from the equation. Therefore, we conclude that the period of the pendulum must not depend on the mass m in our analysis! Thus, we make a big step forward and eliminate m from Eq. (4.8), thus finding that

$$\tau = f(l, g, \alpha) \tag{4.10}$$

Because α is dimensionless, the combination of l and g must have the dimensions of time, and these variables can only appear in Eq. (4.10) as the combination (l/g); thus,

$$\tau = f\left(\frac{l}{g}, \alpha\right) \tag{4.11}$$

Thus, the period (which has dimensions of time) must be proportional to the square root of l/g, or

$$\tau = \left(\frac{l}{g}\right)^{1/2} f(\alpha) \tag{4.12}$$

The dependence of the period on the amplitude remains unknown—because α is a dimensionless variable, and changing it can affect τ without affecting the dimensions of τ.

The implications of this analysis and of Eq. (4.12) are profound. Without resorting to any concepts of physics, we have analyzed the motion of a pendulum and drawn two important conclusions:

1. The period of oscillation is independent of the mass.
2. The period increases in proportion to the square root of the length of the pendulum.

Of course the fundamental laws of physics lead to the same conclusions. They also provide an explicit functional form for the dependence of the period on the amplitude of the oscillations. And so this example is valuable in showing how far dimensional analysis can take us—it tells us about a limit in the number of variables that describe a physical phenomenon, and it provides functional forms relating those variables.

An important general point is that the final result of dimensional analysis almost always needs to be augmented by experiments or a more detailed theory before a full description of the phenomenon is in hand. In the present case, that means either doing experiments to find the form of $f(\alpha)$ or using basic theories of physics to determine it.

THE BUCKINGHAM PI THEOREM

With this example as an introduction, we now make our analysis more formal by stating a limited version of the Buckingham Pi theorem, as follows. In our consideration of a physical phenomenon, we let N_V = the number of variables and N_D = the number of primary dimensions. The theorem tells us that the number of *independent dimensionless groups G* is

$$G = N_V - N_D \tag{4.13}$$

If the P dimensionless groups are a set represented as N_1, N_2, N_3, ... N_P, then it is possible to express any one of them as a function of the others; thus, N_P is a function of N_1, N_2, N_3, ... N_{P-1}.

Next, we develop the meaning of the term "independent dimensionless group." In reading the next sentences, remember that a combination of dimensionless groups is itself dimensionless; for example, one dimensionless group divided by another dimensionless group is yet another dimensionless group. Here is a statement of the theorem: *there is no combination of the dimensionless groups that can be made that would give rise to one of the groups not being included in the combination.* This statement seems formal and awkward, and so we attempt to clarify it below, by example. All of our examples have been chosen for simplicity to be limited to no more than two independent dimensionless groups. Thus, we may specify these expressions as follows:

1. If there is only one independent dimensionless group (N_1), then it cannot depend on another dimensionless group and therefore must be a constant:

$$N_1 = \text{constant} = C_1 \qquad (4.14)$$

2. If there are two independent dimensionless groups (N_1 and N_2), then each must depend on the other:

$$N_2 = f(N_1) \qquad (4.15)$$

To illustrate how to use the Buckingham Pi theorem, we return to the pendulum analysis. Examining Table 4.3, we find that there are five variables and three primary dimensions. Therefore, from Eq. (4.13), we calculate the number of independent dimensionless variables G; it is $5 - 3$:

$$G = 2 \qquad (4.16)$$

We now seek to identify these two dimensionless variables. From what we have already done by way of analysis, we see that one of them must be the amplitude of the oscillation, α. To obtain the second one, we use the following procedure:

1. Consider the variables τ and g, where τ [=] T and g [=] LT^{-2}. To simplify, to find a grouping with only one dimension (we arbitrarily choose L; we could alternatively choose T), we make the combination

$$\tau^2 g \; [=] \; L \qquad (4.17)$$

2. We next simplify further by combining this combination with another variable to identify a dimensionless group; it is sufficient to divide by l; we get
$\tau^2 g/l$, which is a dimensionless group. We could as well have used the inverse, $l/\tau^2 g$.

3. Up to this point, we have not used m, the mass, but we have identified two dimensionless variables. We know that because $G=2$, these two dimensionless variables must be independent of each other.
4. Thus, we define our first dimensionless group α as N_1 and the second one, $\tau^2 g/l$, as N_2, so that the statement

$$N_2 = f(N_1) \tag{4.18}$$

becomes

$$\frac{\tau^2 g}{l} = f(\alpha) \tag{4.19}$$

5. With this, we return to the result developed above, Eq. (4.12), and specify the function as $f_1(\alpha)$,

$$\tau = \left(\frac{l}{g}\right)^{1/2} f_1(\alpha) \tag{4.20}$$

and we thereby recognize that $f_1(\alpha)$ is simply the square root of the arbitrary function, $f(\alpha)$.

Example 4.3 Motion of Water Waves

Problem statement: Use dimensional analysis to guide the analysis of the velocity of a water wave exemplified by a tsunami in deep water.

Solution

We picture an event, such as that which occurred in 2006 in the Indian Ocean, as shown in Figure 4.4.

The wave is initiated in deep water and propagates over long distances. A typical wave motion may be represented by the parameters shown in Figure 4.5. Here, h is the average height of the water above the ocean floor; c is the wave speed; α is the amplitude of the wave; λ is the wavelength or distance between crests; ρ is the density of the water; and g is the acceleration caused by gravity. Table 4.4 is a summary of the relevant quantities and their dimensions.

We have $N_V = 6$ and $N_D = 3$; hence, there are three independent dimensionless variables $(6 - 3)$. As in Example 4.2, because the dimensions of mass appear in only one variable (the density), it is not possible to include the density in any dimensionless group involving our variables. Thus, we eliminate the density as unnecessary for the analysis. With the density eliminated, we are left with $N_V = 5$ and $N_D = 2$ (and realize that there are still three independent dimensionless variables). Because three of our variables (λ, h, and α) each have dimensions of distance, we can easily identify two dimensionless variables, α/h and h/λ. (What other possible pairs can you identify?). A third dimensionless variable may be formed by considering that

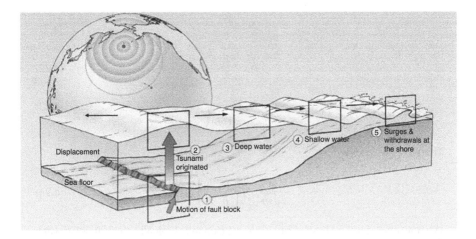

FIGURE 4.4 Various phases of the propagation of a tsunami. *Source*: H. V. Thurman and A. P. Trujillo, "Essentials of Oceanography" (7th edition). Reproduced with permission of publisher (Prentice Hall).

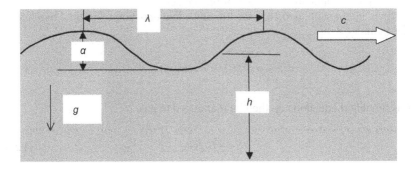

FIGURE 4.5 Wave motion in water of depth h.

TABLE 4.4
Variables for Analysis of Wave Propagation

Name of Quantity	Symbol	Dimensions
Wave speed	c	LT^{-1}
Wavelength	λ	L
Mean depth of water	h	L
Gravitational acceleration	g	LT^{-2}
Amplitude of wave	α	L
Density of water	ρ	ML^{-3}

$$\frac{c}{\sqrt{g}} [=] \sqrt{L}$$

so that if we divide this by $h^{1/2}$, we get a third dimensionless variable, N_3. Thus, we have

$$N_1 = \frac{\alpha}{h} \tag{4.21}$$

$$N_2 = \frac{h}{\lambda} \tag{4.22}$$

$$N_3 = \frac{c}{(gh)^{\frac{1}{2}}} \tag{4.23}$$

Therefore, we can write

$$\frac{c}{\sqrt{gh}} = f\left(\frac{\alpha}{h}, \frac{h}{\lambda}\right) \tag{4.24}$$

We now consider the physical situation of a tsunami. Typically, these involve water that is 2–4 km deep. The wavelengths are hundreds of kilometers. The amplitudes of these waves are small relative to either h or λ. Using these values, we see that both α/h and h/λ are orders of magnitude smaller than unity. Thus, we are tempted to consider these dimensionless groups to be so small as to be negligible—that is, we consider the limit in which they approach zero.

Then, in our simplified analysis, the preceding equation takes on a much simpler form:

$$\frac{c}{\sqrt{gh}} = f(0,0) = \text{constant} = K \tag{4.25}$$

This simplified equation can then be rearranged to give

$$c = K\sqrt{gh} \tag{4.26}$$

A detailed theory of water waves shows that the constant K has the value of one. Experiments show that the wave speed varies with the square root of the height, in agreement with Eq. (4.26). This result implies that the deeper the water (the greater the value of h), the faster the wave propagates (the greater the value of c).

Suppose, for example, that $h = 4,000$ m, then $c = 200$ m/s (or 720 km/h). And if $h = 250$ m, then $c = 50$ m/s (or 180 km/h). These estimates are in line with observations; indeed, these formulas have been used with computer models of the ocean floor and found to be excellent estimates for the wave propagation speeds in the central part of the ocean.

TANK DRAINING

Next, to connect dimensional analysis to a familiar topic, we apply it to tank draining (Figure 4.6). We use dimensional analysis to determine the relationship between the height of water in the tank and the volume flow rate. Thus, we will check what we did in Chapter 2. We consider the draining tank from the standpoint of determining the volume flow rate through an orifice, with a flow that results from a particular pressure drop across the orifice. That is, we assume that we are given an orifice of cross-sectional area A_o. The pressure drop across the orifice is ΔP, and

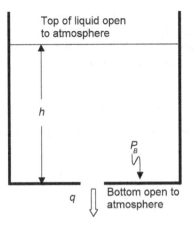

Top of liquid open
to atmosphere

h

P_B

q

Bottom open to
atmosphere

FIGURE 4.6   Schematic representation of draining tank.

TABLE 4.5
Summary of Variables for Dimensional
Analysis of Flow across the Orifice

Name of Quantity	Variable	Dimensions
Volume flow rate of fluid	q	L^3T^{-1}
Pressure drop across orifice	ΔP	$ML^{-1}T^{-2}$
Density of flowing fluid	ρ	ML^{-3}
Cross-sectional area of orifice	A_o	L^2

we let ρ be the mass density of the fluid (recall that we used this symbol above to represent molar density rather than mass density). Our goal now is to determine how q, the volume flow rate of the fluid, depends on these variables, which are summarized in Table 4.5.

Before proceeding, we consider the meaning of ΔP. This is the difference between the pressure in the fluid at the bottom of the tank and the pressure in the fluid just outside of the tank. The pressure outside of the tank is the atmospheric pressure, that is, the pressure that we experience around us.

The atmospheric pressure depends, for example, on the temperature and the elevation. To visualize how elevation affects the air pressure, think of how a plastic container that you close in the valley before driving to the mountains will bulge out as you drive up to higher elevations. The bulging results because the pressure in the container stays the same (provided that the temperature stays the same), but the pressure outside the container decreases with increasing elevation.

Suppose that we had a way of controlling the pressure outside the tank to make it different from atmospheric pressure; if we made this pressure equal to the pressure in the fluid at bottom of the tank, then there would be no flow out of the tank. If the pressure difference were small, the flow rate of liquid would be small, and the greater the pressure difference, the faster the flow. This is the broad idea that we will capture in our dimensional analysis.

According to Table 4.5, the number of variables is four and the number of primary dimensions is three. Hence, by the Buckingham Pi theorem, the number of independent dimensionless groups is $4 - 3 = 1$. We begin by combining the pressure drop and the density to eliminate the dimension of mass:

$$\frac{\Delta P}{\rho} [=] \frac{ML^{-1}T^{-2}}{ML^{-3}} [=] L^2 T^{-2} \tag{4.27}$$

Next we use the volume flow rate to eliminate the time dimension:

$$\frac{\Delta P}{\rho q^2} [=] \frac{L^2 T^{-2}}{L^6 T^{-2}} [=] L^{-4} \tag{4.28}$$

Now, we can multiply by the square of A_o to obtain a dimensionless quantity:

$$N_1 = \frac{\Delta P A_o^2}{\rho q^2} \tag{4.29}$$

As noted above, we can now set $N_1 = $ constant, which we call $1/C^2$.

$$\frac{1}{C^2} = \frac{\Delta P A_o^2}{\rho q^2} \tag{4.30}$$

By rearranging this equation, we see that

$$q^2 = C^2 \frac{\Delta P A_o^2}{\rho} \tag{4.31}$$

or

$$q = C A_o \sqrt{\frac{\Delta P}{\rho}} \tag{4.32}$$

The next step is to develop a relationship between the pressure drop and the height of the liquid in the tank. The top of the tank is open to the atmosphere, and so the pressure there is the atmospheric pressure, denoted by P_{atm}. Therefore, the total force acting on the fluid at the top of the tank is $P_{atm} A_T$ (do not confuse A_T, the cross-sectional area of the tank, with A_o, the cross-sectional area of the orifice).

The pressure in the fluid at the bottom of the tank is greater than that in fluid at the top of the tank because of the force of gravity exerted on the fluid. The volume of liquid in the tank, V_T, is given by

$$V_T = A_T h \tag{4.33}$$

(recall that A_T is the cross-sectional area of the tank and h is the height of liquid in the tank (above the orifice)). The mass of liquid in the tank, M_T, is

$$M_T = \rho V_T \tag{4.34}$$

The force associated with this mass, or the weight of the liquid, is obtained by multiplying the mass by the gravitational acceleration, or, $F = ma$, where $a = g$:

$$W_T = gM_T \tag{4.35}$$

The pressure is the force per unit area. Therefore, the pressure on the bottom of the tank resulting from the gravitational pull on the water, P_T, is W_T/A_T. Combining Eqs. (4.33)–(4.35), we find

$$P_T = \rho gh \tag{4.36}$$

The total pressure on the bottom of the tank, P_B, has two contributions, the pressure attributed to the liquid in the tank and the atmospheric pressure that acts on the liquid at the top of the tank, P_{atm}. These are additive:

$$P_B = P_{atm} + P_T \tag{4.37}$$

We would have arrived at the same result by considering that the total force (weight) acting on the bottom of the tank is the sum of the weight of fluid and the atmospheric pressure, $P_{atm}A_T$ and gM_T.

We seek to find ΔP. This is the difference in pressure between the bottom of the tank and the outside. The outside pressure is atmospheric pressure, P_{atm}, or

$$\Delta P = P_B - P_{atm} = P_T \tag{4.38}$$

Combining this with Eqs. (4.38), (4.36), and (4.32), we obtain

$$q = CA_o \sqrt{gh} \tag{4.39}$$

This result matches what we found before. It reinforces all of our previous findings that led us to conclude that the flow rate q is proportional to $h^{1/2}$. We can combine $CA_o g^{1/2}$ into a single empirical constant, C_1, and write, again in agreement with what we found in Chapter 2,

$$q = C_1 \sqrt{h} \tag{4.40}$$

To summarize, to this point we have shown that for liquid in a tank draining under the influence of gravity, the volume flow rate is proportional to the square root of the height of liquid above the orifice.

We have now obtained this result by several different means. First, we analyzed experimental results and used physical reasoning to find an equation that provided a good fit of the data; then we considered the principle of conservation of mass in our analysis; and now we have obtained the same result by using dimensional analysis.

The agreement of the results emerging from all these approaches is of course not an accident. A standard fluid mechanical treatment of the draining tank on the basis of fundamental physics leads to the same result.

TABLE 4.6

Summary of Variables for Dimensional Analysis of Tank Draining

Name of Quantity	Variable	Dimensions
Volume flow rate of fluid	q	L^3T^{-1}
Gravitational acceleration	g	LT^{-2}
Density of fluid	ρ	ML^{-3}
Cross-sectional area of orifice	A_o	L^2
Height of liquid above orifice	h	L

The tank draining analysis can be done in yet another way. In the example above using dimensional analysis, we needed to develop an equation to describe the pressure drop across the orifice. In that example, we were deriving the form of a constitutive equation for the flow through the orifice. Alternatively, we can consider our system as the entire tank and develop our relationship between q and h directly. Proceeding to do that, we identify the relevant variables listed in Table 4.6.

The number of dimensions is three, and the number of variables is five. Therefore, the number of independent dimensionless variables is two. As in the foregoing example, it is possible to eliminate the density as a variable, because no other variable contains the dimension M. This step leaves us with two dimensions and four variables.

By examination, we can identify the dimensionless variable

$$N_1 = \frac{A_o}{h^2} \tag{4.41}$$

To identify N_2, we consider the combination

$$\frac{q}{\sqrt{g}} [=] \frac{\dfrac{L^3}{T}}{\dfrac{L^{1/2}}{T}} [=] L^{\frac{5}{2}} \tag{4.42}$$

This can be made dimensionless by using A_o and h, as follows:

$$N_2 = \frac{q}{A_o\sqrt{gh}} [=] \frac{L^{\frac{5}{2}}}{L^2 L^{1/2}} \tag{4.43}$$

Now, we know that according to the Buckingham Pi theorem,

$$N_2 = f(N_1) \tag{4.44}$$

In the limit as the cross-sectional area of the orifice becomes small relative to the square of the height of the liquid, N_1 becomes small, approaching a limiting value of zero. Thus, $f(N_1)$ tends to a constant value, and we are left with the result that $N_2 =$ constant, and we get the result found in the preceding development.

CORRELATIONS OF DIMENSIONLESS GROUPS

Many complex physical phenomena involving fluid flow, transport of heat, and transport of mass are described compactly with relationships among dimensionless groups. Discovery of these correlations is guided by dimensional analysis, and values of the parameters are determined by experiment. The correlations are broad generalizations that allow many predictions that are valuable for design. It is usually not a good idea to extrapolate the correlations, as they are based on limited experimental results.

We now consider an example of such a correlation involving transport of molecules to the surfaces of particles in a fixed bed through which fluid flows. The correlation also accounts for transport of molecules from the particle surfaces into the flowing fluid (they might be formed by dissolving the solid). The molecules are transported in the fluid by both random molecular motion (diffusion) and convection. The process is influenced by the fluid properties and also the particle sizes and shapes.

It is important to recognize that flow in a packed bed is influenced by the particles in it, and so the Reynolds number is defined differently from that characterizing flow in a pipe without particles. Dimensionless groups that have been found to influence the rates of transport of the molecules to or from the particle surfaces include the following:

1. A dimensionless term called the J-factor, J_D, defined as $k_c u N_{Sc}^{2/3}$, where N_{Sc} is the Schmidt number, $\mu D/\rho$, where μ is the viscosity of the fluid, D is the diffusion coefficient of the compound being transported in the fluid, and ρ is the mass density of the fluid. The term k_c is called a mass transfer coefficient, which accounts for transport of molecules by diffusion and convection. The term u is called the superficial velocity, G/ρ, where G is the superficial mass flow rate, based on the cross-sectional area of the tube, without account taken of the particles in it—so that G is the mass flow rate per cross-sectional area of the tube. G/ρ has SI units of m s^{-1}. (Can you check what the SI units of G are?)

2. The modified Reynolds number; this is defined as $d_p G/\mu$, where μ is the fluid viscosity, G is as defined above, and d_p is the equivalent diameter of the particle, that is, the diameter a sphere of the same volume as the particle, which is defined as $[(6/\pi)(\text{particle volume})]^{1/3}$. When there is a distribution of particle sizes and shapes, as is typical, average values are used. Can you confirm that this modified Reynolds number is dimensionless?

3. Another dimensionless parameter is needed to account for the presence of the particles in the tube through which the fluid flows: this is Φ, the void fraction of the fixed bed, that is, the fraction of the bed that is not occupied by solid and instead is occupied by fluid (this is also called the poro sity of the bed; typical values are roughly 0.5).

With these parameters, a correlation relating the J-factor and the modified Reynolds number has been determined experimentally for gases flowing through packed beds, as shown in Figure 4.7. As shown by the many data points on the figure, the relationship on a log–log plot is nearly linear over orders of magnitude of the modified Reynolds number.

This is one of the many examples of such correlations used by chemical engineers.

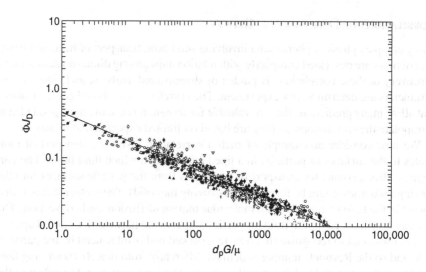

FIGURE 4.7 Correlation for prediction of mass transfer coefficients for gases in packed beds. The terms are defined in the text. *Source*: Reproduced with permission from P. N. Dwidevi and S. S. Upadhyay, Particle-Fluid Mass Transfer in Fixed and Fluidized Beds, *Ind. Eng. Chem. Process Des. Dev.*, **1977**, *16*, 157.

RECAP AND REVIEW QUESTIONS

The requirement of dimensional homogeneity provides a check of equations—any equation in which not every term has the same dimensions is incorrect. The Buckingham Pi theorem underlies the method of dimensional analysis, whereby we identify the number of independent dimensionless variables to represent a physical phenomenon and thereby gain valuable insight into the form of an equation describing the phenomenon. If N_V=the number of variables describing the phenomenon and N_D=the number of primary dimensions, then the theorem tells us that the number of independent dimensionless groups is $G=N_V-N_D$. Chemical engineers use many correlations determined experimentally involving dimensionless groups. Will you remember to include units each time you are checking an equation as a check on its internal consistency?

PROBLEMS

4.1. An equation for a non-ideal gas is written as follows:

$$\left(P + \frac{N^2 a}{V^2}\right)(V - Nb) = NRT$$

    A. Show that in a limiting case, this equation reduces to the ideal gas law, and determine what the symbols P, V, and N represent.

    B. What are the dimensions of the constants a and b, and what are their units in the SI system?

    C. Use the equation to determine the units of the gas constant R in the SI system.

    D. Check that the criterion of dimensional homogeneity is met by this equation.

4.2. Determine the SI units of the parameters A_0, b, B_0, and C_0 in the BWR equation of Example 4.1.

4.3. Comment on whether the BWR equation in Example 4.1 was developed on the basis of dimensional analysis. Explain your reasoning.

4.4. An equation describing the motion of molecules in an ideal gas is the following:

$$u = \left(\frac{8RT}{M\pi} \right)$$

Here R and T are the gas constant and the absolute temperature, respectively, and M is the molecular mass of the gas. Figure out what u is and what dimensions it has. Does it make physical sense that u increases as the temperature increases and as the mass of the gas molecule decreases?

4.5. Check the following equations, which pertain to tank draining, for dimensional homogeneity and identify any that might be incorrect just on the basis of a lack of consistency of the dimensions:

$$q = kh^n$$

$$\left(\frac{dh}{h^n} \right) = -\left(\frac{k}{A} \right) dt$$

$$h = h_0 \left\{ 1 - \left[\frac{k(1-n)}{h_0^{(1-n)}} \right] t \right\}^{\frac{1}{(1-n)}}$$

4.6. A textbook equation accounting for transport of molecules from a flowing fluid to the surfaces of solid particles and then further into pores of the solid particles is the following:

$$\left(\frac{D_{eA}}{L} \right) \left(\frac{dC_A}{dL'} \right) = k_g \left(C_{ab} - C_A \right)$$

In this equation, L and L' have dimensions of length, the C terms are concentrations, and D_{eA} is a diffusion coefficient. What are the dimensions of k_g? What are its units in the SI system?

4.7. A textbook equation that describes chemical reaction taking place in a porous solid particle is the following:

$$\frac{R_f}{R} = \left\{ \frac{\tanh\left[\varphi_{pl} (1 - p) \right]}{\tanh \varphi_{pl}} \right\} \left\{ \frac{1}{\left(1 + \varphi_{pl} p \right)} \right\}$$

Find out what a hyperbolic tangent (tanh) is and determine which terms in this equation are dimensionless, which are not, and which may or may not be. Explain your reasoning.

4.8. A well-known correlation of dimensionless groups expresses a term called the friction factor, which accounts for pressure drop in pipe flow, as a function of Reynolds number. This plot illustrates points made in this chapter about laminar and turbulent flow. Do some research and find the plot and reconcile it with the results stated in this chapter. What new dimensionless group related to the roughness of the pipe surface appears in the plot?

4.9.   In a dimensional analysis of the performance of a rotameter with a tapered inside tube diameter and a spherical float, what dimensionless groups representing dimensions of the apparatus do you consider to be essential? What other dimensionless groups do you anticipate to be essential in the overall analysis?

4.10.   The flow of blood in the human body is governed by several dimensionless groups. Consider the dimensionless number formed from the frequency of the pumping of the heart (which creates a periodic flow), f [=] T^{-1}; the diameter of the blood vessel is approximated as a constant, d; the density of the blood is ρ; and the viscosity of the blood is μ [=] $ML^{-1}T^{-1}$. What is the dimensionless number associated with these four parameters?

4.11.   Use dimensional analysis to consider the problem of a falling mass released from rest under the influence of gravity. Determine a relationship among the distance traveled, s, the acceleration of gravity, g, and the time, t. Show that dimensional analysis leads to the conclusion that $s/gt^2 = K$, where K is a constant.

4.12.   Consider flow of a fluid around a circular cylinder that is infinitely long. At some Reynolds numbers that are sufficiently low, the flow is described as creeping flow.

    A.   Check out the meaning of this term and find out how the Reynolds number is defined and what range of Reynolds numbers corresponds to creeping flow. Also find out what range of Reynolds numbers corresponds to turbulent flow and provide an order-of-magnitude ratio of the lowest Reynolds number corresponding to turbulent flow to the highest Reynolds number corresponding to creeping flow.

    B.   Identify a realistic situation in which creeping flow takes place.

4.13.   Figure out the dimensions of the term k_c used to introduce Figure 4.7.

4.14.   Consider the draining tank for which we have done dimensional analysis of the draining process, and propose additional terms to consider in a dimensional analysis when the water flow from the tank is dropwise and not continuous. Think about what could be observed with a strobe light and a camera to characterize the discontinuous flowing stream.

4.15.   Find an equation to represent the data in Figure 4.7.

4.16.   The authors P. N. Dwidevi and S. S. Upadhyay, who reported Figure 4.7 (and whose work appears in P. N. Dwidevi and S. S. Upadhyay, Particle-fluid mass transfer in fixed and fluidized beds, *Ind. Eng. Chem. Process Des. Dev.*, **1977**, *16*, 157–165), also reported an equation to represent data characterizing mass transfer for both gases and liquids in both fixed and fluidized beds. This equation is thus more general than the correlation shown in Figure 4.7, and would not be expected to agree with it exactly. The equation is the following, where N_{Re} is $d_p G/\mu$ as that term is used in Figure 4.7, and the terms are those stated in the text accompanying Figure 4.7:

$$\Phi J_D = \frac{0.765}{\left(N_{Re}\right)^{0.82}} + \frac{0.365}{\left(N_{Re}\right)^{0.386}}$$

Make a plot to show how well this correlation agrees with that of Figure 4.7.

4.17. In the book *Chemical Engineering*, Volume 1, Third Edition, by J. M. Coulson and J. F. Richardson (Pergamon Press, Oxford, 1977), there is an introduction to dimensional analysis in which the authors (Example 1.3, pp. 8–10) use dimensional analysis to consider the pressure difference between two ends of a pipe in which a fluid is flowing as a function of the following variables: the pipe diameter, the pipe length, the fluid viscosity, and the fluid density. They show how the analysis leads to the identification of the Reynolds number as a criterion for the type of flow. The approach taken by these authors is complementary to that presented here. Find the presentation by Coulson and Richardson and summarize it briefly, along with the important lessons.

5 Chemical Reactors and Chemical Reactions

ROADMAP

In this chapter, we add a significant new component to our analysis: we bring in chemical change. Now, we go beyond the mass balance as an approach to setting up an analysis, as we take account of the change of one or more components into other components as a result of chemical reaction. The approach is to write a balance equation on molecules and to recognize that molecules, in contrast to mass, are not conserved—they are interconverted in chemical reactions. And so the balance equations we use now are of a new kind—they are not conservation equations, but rather are equations that are like mass balance equations, except that they have a new term, called a generation term, which accounts for the formation (or depletion) of each kind of molecule by chemical reaction. Now the equations are written not in terms of mass per unit time but rather molecules per unit time. By applying these molecule balance equations, we do not abandon the principle of conservation of mass; instead, we just add another kind of equation to the development in addition to the mass balance. We start with a general molecule balance and simplify it for three special cases, corresponding to three ideal chemical reactors—reactors are the vessels in which chemical reactions occur. The molecule balance equations describe the reactors and not the reactions. Later in this chapter, we introduce equations describing the reactions (these are constitutive relationships), and then we combine the equations describing the reactors with the equations describing the reactions, and the results allow us to predict the performance of reactors for particular reactions—and design the reactors.

ACCOUNTING FOR CHEMICAL REACTIONS

In the quantitative developments so far, we have restricted ourselves to physical processes such as flows rather than processes involving chemical change. Now we include chemical reactions as well as flows. By adding reactions to our analysis, we enter territory that is most uniquely associated with chemical engineering, biochemical engineering, and civil and environmental engineering, as well as chemistry. Engineers in these fields are concerned with chemical change and all the equipment and processes used to carry out chemical change. Biochemical engineers deal with production of chemicals from biological raw materials, such as yeast and wood. Environmental engineers deal with purification of waste water, for example. Chemical engineers also deal with these subjects, and also with the production of chemicals and fuels from many natural resources, such as natural gas, petroleum, coal, air, and water, as well as biomass, such as wood. Thus, all these engineers

DOI: 10.1201/9781003429944-5

deal with pipes, pumps, and valves in systems that deliver reactants to the vessels in which they are converted; the engineers also deal with the devices used to purify the products of chemical change and deliver them for further processing or use. And, as we see in this chapter, they deal with the vessels in which chemical change takes place. These are called reactors. We now proceed to analyze the performance of several simple (ideal) reactors.

The accounting statement that represents how much of each reactant is consumed and how much of each product is formed when the reaction proceeds to completion is the stoichiometry of the reaction. This is the most basic information characterizing a chemical reaction.

Consider the example of the ammonia synthesis, an extremely important industrial reaction:

$$3H_2 + N_2 \rightarrow 2NH_3 \tag{5.1}$$

NH_3 is generated; H_2 and N_2 are consumed. Mass is of course still conserved (and mass balance equations are no less valid or useful than when no reaction occurs), but when we focus on the chemical change and the reactor in which it occurs, we need a balance equation that has generation terms accounting for the chemical reactions. In the ammonia synthesis reaction as written in Eq. (5.1), the stoichiometric coefficients are 3, 1, and 2 for H_2, N_2, and NH_3, respectively; often we write these as −3, −1, and +2, respectively, to account with the negative signs for the fact that the H_2 and N_2 are consumed and with the positive sign for the fact that NH_3 is formed. Stoichiometry is about the proportions of reactants and products, and so we could replace the stoichiometric coefficients −3, −1, and +2 with −1, −1/3, and +2/3, for example (among others), without changing the meaning of the stoichiometric equation.

REACTION RATES

The rate of a reaction is a measure of how fast it proceeds, and the relative rates of consumption of reactants and formation of products are related to each other by the reaction stoichiometry. In the ammonia synthesis, two mols of ammonia are formed for each mol of N_2 that is consumed, for example. These ideas are illustrated in the following example.

Example 5.1   Determining a Reaction Rate from Candle-Burning Data

Problem statement: In the burning of a candle, wax reacts with O_2 to make CO_2 and water (and side products such as CO and soot, the rates of formation of which we may consider to be approximately negligible with respect to the rates of formation of CO_2 and water). Use the data in Figure 5.1 to determine an approximate equation for the height of a burning candle that is cylindrical with a diameter of 2.8 cm as a function of time. No wax flowed from the candle as it burned. Show how these data determine an approximate rate of formation of CO_2 and determine an equation for this rate. State the assumptions used in the analysis.

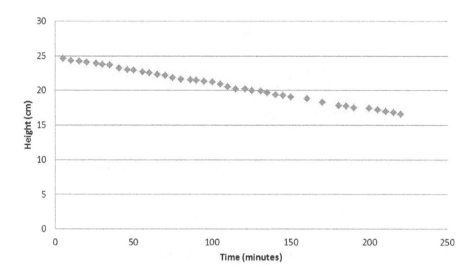

FIGURE 5.1 Data that determine rate of burning of a candle.

Solution

The data in Figure 5.1 demonstrate a nearly linear relationship between the candle height and time. This plot therefore demonstrates a nearly constant rate of burning, thus a nearly constant rate of formation of CO_2 and water from wax, provided that we can neglect side reactions. We place a straight line through the data and find the following equation to represent them (where h is the candle height in cm and t is the time in min): $h = -0.0373t + 24.8$. To determine the rate of formation of CO_2, we need some information about the candle wax. Wax is an alkane (a paraffin), with the composition C_nH_{2n+2}, with n typically being in the range of 20–40. As an approximation, let us assume that the wax is $C_{20}H_{42}$ (with a molecular mass, often called a molecular weight, of 282.4 g/mol). We look up the density of solid $C_{20}H_{42}$ (n-eicosane), finding it to be 0.932 g/cm³. The stoichiometry of the wax burning is as follows: $C_{20}H_{42} + 30.5\ O_2 \rightarrow 20\ CO_2 + 21\ H_2O$. To determine the rate of production of CO_2, we need to determine the rate of consumption of the wax and use the stoichiometry. This value is determined by the slope of the line representing the data in Figure 5.1, and we write $dh/dt = -0.037$ cm/min.

This information (with the candle diameter) determines the rate of change of volume (V) of the candle, from which we calculate the rate of change of mass (M) (by using the density) and then calculate the rate of change of the number of mols (from the molecular weight). Then we use the stoichiometry to determine the rate of formation of CO_2, which we call dN/dt, where N is the number of mols of CO_2. Thus,

$$\frac{dV}{dt} = \frac{dh}{dt} \times \text{cross-sectional area}$$

$$\rho \times \frac{dV}{dt} \rightarrow \frac{dM_{C_{20}H_{42}}}{dt} \times \left(\frac{1}{MW_{C_{20}H_{42}}}\right) \rightarrow \frac{dN_{C_{20}H_{42}}}{dt} \times \text{ratio of stoichiometric coefficients}$$

$$\rightarrow \frac{dN_{CO_2}}{dt}$$

$$\text{Candle cross - sectional area } A = \pi \times \left(\frac{2.8}{2}\right)^2 = 5.16 \text{ cm}^2$$

$$\frac{dV}{dt} = A \times \frac{dh}{dt} = 5.16 \times (-0.0373) = -0.23 \text{ cm}^3/\text{min}$$

$$\frac{dM_{C_{20}H_{42}}}{dt} = \rho \times \frac{dV}{dt} = 0.932 \times (-0.23) = -0.21 \text{ g/min}$$

$$\frac{dN_{C_{20}H_{42}}}{dt} = \frac{dM_{C_{20}H_{42}}}{dt} \times \left(\frac{1}{MW}\right) = -0.214 \times \left(\frac{1}{282.4}\right) = -7.4 \times 10^{-4} \text{ mol/min}$$

$$\frac{dN_{CO_2}}{dt} = \frac{dN_{C_{20}H_{42}}}{dt} \times \text{ratio of stoichiometric coefficients}$$

$$= -7.4 \times 10^{-4} \times \left(\frac{20}{-1}\right) = 0.015 \text{ mol/min}$$

$$\frac{dN_{CO_2}}{dt} = 0.015 \text{ mol/min} \times \left(\frac{1 \text{ min}}{60 \text{ s}}\right) = 2.5 \times 10^{-4} \text{ mol/s};$$

this is the rate of CO_2 formation.

BATCH AND FLOW SYSTEMS

To proceed with the analysis of reactors, we need some new terms. We define a *batch system* as one into which and from which there is no flow; there may be flow within the system, but nothing leaves or enters it. The opposite of a batch system is a *flow system*. We have considered numerous examples of flow systems, beginning with the simple draining tank. Some chemical reactors are batch reactors, and others are flow reactors. In describing a flow reactor, we refer to a feed stream as one that enters the reactor and a product stream as one that leaves it. If reaction occurs in the reactor, the feed and product streams have different compositions. Usually, only a fraction of the reactants are converted, and so reactants are present along with products in the product stream. The product stream may flow to a separations device to be purified, giving a stream with a higher concentration of product and, as we understand from our analysis of multicomponent mass balances, a stream or streams with higher or lower concentrations of other compounds, such as the unconverted reactants. Some streams may be sent back to the reactor (recycled) to give the reactants more opportunity to be converted.

Many reactors used in industry for large-scale production are flow reactors, allowing the operating efficiency of continuous feeding of reactants and continuous formation of products. Batch reactors are also used for industrial production—usually on a small scale—but they are most common in laboratories where scientists and engineers explore new chemistry. Batch reactors are convenient to use and allow experimentation with small quantities of chemicals.

STEADY-STATE AND TRANSIENT PROCESSES

Recall the ideas of steady-state and transient processes that were introduced in Chapter 1: When a process operates so that any property of the contents in it is characterized by a lack of time dependence, the system operates at *steady state* (also called stationary state). For example, a tank that is being filled with a liquid at exactly the rate at which liquid drains from the tank is operating at steady state, provided that properties such as temperature, pressure, and composition of the tank contents are independent of time.

When any of these properties or, for example, the mass or volume of the contents of the tank changes with respect to time, the process is called a *transient* (or unsteady-state) process. Many large-scale industrial processes operate at nearly steady state most of the time, but when they are started up or shut down, or when there is an inadvertent upset such as caused by a power failure or a plug in the flow system, they operate transiently. A transient operation of a large-scale industrial process typically approaches a steady state over time.

Transient processes are common. Consider a cow as a chemical reactor. The feeds entering the cow and the products leaving it are intermittent and depend on the time of day, and the reactions occurring in the cow's digestive tract consequently occur transiently as well, depending not only on the intakes and effluents but also on other physiological processes, such as whether the cow is walking or sleeping.

ANALYSIS OF CHEMICAL REACTORS AND THE ANALOGY
WITH AN INTEREST-BEARING BANK ACCOUNT

Earlier, we considered a no-fee, no-interest bank account to introduce the concept of conservation of mass. Written schematically, the equation was the following:

$$\text{in} - \text{out} = \text{accumulation}$$
$$\frac{\$}{\text{time}} \quad \frac{\$}{\text{time}} \quad \frac{\$}{\text{time}} \tag{5.2}$$

This balance equation is a conservation equation, analogous to a mass balance equation (dollars are conserved; and, analogously, mass is conserved).

Now, to introduce chemical reactors, consider an interest-bearing bank account (we assume for simplicity that there are no fees). Because money in this account generates interest, there is a new term in the balance equation, called the generation term:

$$\begin{array}{ccccccc} \text{in} & - & \text{out} & + & \text{interest} & = & \text{accumulation} \\ \dfrac{\$}{\text{time}} & & \dfrac{\$}{\text{time}} & & \dfrac{\$}{\text{time}} & & \dfrac{\$}{\text{time}} \end{array}$$

(5.3)

This is a generation term.

Money is generated by interest.

If there were fees associated with the account, there would be yet another term:

$$\text{in} - \text{out} + \text{interest} - \text{fees} = \text{accumulation}$$

(5.4)

This is essentially a

(negative) generation term.

Money is lost to fees.

The preceding two equations are balance equations but *not conservation equations*—money is not conserved in these accounts. The presence of one or more generation terms sets these equations apart from conservation equations.

An essential point for analysis is that balance equations may be extremely useful starting points, even when they are not conservation equations. We will use a balance equation with a generation term to account for chemical reactions. In a chemical reaction, some kinds of molecules are converted into other kinds of molecules as chemical bonds break and/or form. Thus, reactants are consumed (negative generation), and products are formed (positive generation).

ANALYSIS OF CHEMICAL REACTORS: THE GENERAL MOL BALANCE EQUATION

In a mass balance equation, we used terms containing mass units, such as kg/s. In the balance equations we will use now, we account for molecules. We will write molecule balances. By convention, we write the equations for mols (moles) instead of molecules (a mol of a particular compound is Avogadro's number (6.023×10^{23}) molecules of that compound). Remember that molecules are not conserved; they can be transformed into other kinds of molecules.

To proceed with our analysis of chemical reactors, we use two kinds of equations, which are entirely separate from each other. *One kind of equation is a balance equation written for* one of the kinds of *molecules that are in the reactor and may enter and/or leave a reactor—this is called a mol balance equation.* The molecules entering and/or leaving could be reactants, products, solvents, or impurities, but the equations will be most useful when they are reactants or products—then there will be a generation term in the equation. The other kind of equation, complementing the mol balance equation, is a *constitutive relationship describing the reaction*(s) occurring in the reactor.

Below, we summarize important points about these two kinds of equations. It is essential to keep in mind that they are separate from each other—one describing the reactor and the other describing the reaction. As we shall see, they are linked to each other by a common term.

For the *system* (which may be the reactor—where the reaction takes place— or part of the reactor), we use a balance equation on molecules (mols). The rate of the reaction appears in the balance equation. This equation is a mol balance equation.

For the *reaction*, we use a constitutive relationship that shows how the reaction rate per unit volume r_i depends on the variables that affect it. These variables are primarily the *concentrations* of reactants (and possibly products) and the *temperature*. Here, r_i = rate of formation of compound i in a fluid phase, with units of, say, $\dfrac{\text{mol}}{\text{L} \times \text{s}}$.

Experimental results show that at a given temperature,

$$r_i = k \prod_j C_j^{\alpha_j}$$

and so, for example,
$r_j = kC_i$ (k is a constant)
or $r_j = k'C_i^2$
or $r_j = k''C_iC_k$

At this stage of our analysis, we simply leave it as intuitively clear that the rate of a reaction depends on temperature and on the concentrations of molecules involved in the reaction. We return to the constitutive relationships in some detail later. But first we pursue the mol balance equation.

Now, consider the reactor or part of the reactor to be the system (control volume). We write a general balance equation to account for flow into the system, flow out of the system, generation by chemical reaction in the system, and accumulation in the system. The equation is written for any particular species j (j could be a reactant or a product, say nitrogen or ammonia, for example).

$$F_{j0} \quad - \quad F_j \quad + \quad G_j \quad = \quad \frac{dN_j}{dt}$$

↑	↑	↑	↑
mols of j	mols of j	rate of generation	rate of
flowing	flowing out	of j in system by	accumulation
into system	of system	chemical reaction	of j in system
$\dfrac{\text{mol}}{\text{time}}$	$\dfrac{\text{mol}}{\text{time}}$	$\dfrac{\text{mol}}{\text{time}}$	$\dfrac{\text{mol}}{\text{time}}$
−	−	−	−
flow term	flow term	reaction term	accumulation
		(generation term)	term

(5.5)

To proceed, we need to relate the generation term G_j to the rate of the chemical reaction per unit volume, r_j; making this connection will allow us to link the analysis with the constitutive relationship. Consider the system shown in Figure 5.2, and assume that the contents are a single-phase fluid, so that we can consider it to be a continuum and use the methods of calculus as we have in the preceding chapters. Be sure to notice that the subscript 0 refers to the inlet of the system; this usage appears repeatedly in the following developments.

$$\text{Average rate in fluid element } \#1 = \bar{r}_{j1} \left(\frac{\text{mol}}{\text{volume} \times \text{time}} \right)$$

$\bar{r}_{jn}$ = average rate in fluid element n

We emphasize that the rate may vary with position in the system; in other words, $\bar{r}_{jn}$, the average rate in fluid element n, may be different from that in fluid element m. We also emphasize that we have made no assumption (yet) about whether the system operates at steady state or not. If it does not, then $\bar{r}_{jn}$ is a function of time.

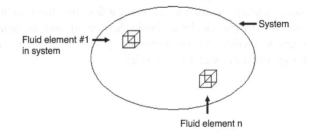

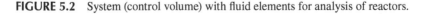

FIGURE 5.2   System (control volume) with fluid elements for analysis of reactors.

We find G_j for the whole system by summing up the generation terms for all the fluid elements:

$$G_j = \Delta G_{j1} + \Delta G_{j2} + ... \qquad (5.6)$$

$$G_j = \sum_{i=1}^{M} \Delta G_{ji}, \quad M = \text{total number of fluid elements} \qquad (5.7)$$

$$G_j = \sum_{i=1}^{M} \bar{r}_{ji} \Delta V_i, \quad \Delta V_i = \text{volume of fluid element } i \qquad (5.8)$$

Now, using calculus, consider the limit as $M \rightarrow \infty$ and $\Delta V_i \rightarrow 0$. Then $G_j = \int^{V} r_j \, dV$,

the integral over the whole volume of the system. Now r_j is a point value and not an average value, because in the limit we have considered the case in which the volume of each fluid element approaches zero.

Substituting this result into Eq. (5.5), we now have the general mol balance equation valid for any system:

$$F_{j0} - F_j + \int^{V} r_j \, dV = \frac{dN_j}{dt} \qquad (5.9)$$

To proceed, we consider *three special cases* of this general balance equation, and each corresponds to one of three *ideal reactors*. Notice that we have not yet specified the system except in a very general way.

THE STEADY-STATE BACKMIX REACTOR

Our first limiting case is the ideal reactor called the *steady-state backmix reactor*, represented schematically in Figure 5.3. We define the system as the volume of the whole reactor and make two assumptions:

1. *Steady state* (this implies that there is no accumulation in the system, and the right-hand term in Eq. (5.9) goes to zero).
2. *Perfect mixing* (backmixing) of the contents of the system, which implies that the *composition is the same everywhere* in the system, and so is the temperature (and the pressure, and the concentration of each species present). Because the reaction rate r_j depends on concentrations and temperature, this rate is therefore the same everywhere in the reactor (and a constant value).

Thus, terms in the general mol balance equation (Eq. 5.9) become simplified as follows:

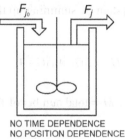

FIGURE 5.3   Schematic representation of a backmix flow reactor.

$$\int^{v} r_j \, dV \rightarrow r_j V \text{ and } \frac{dN_j}{dt} \rightarrow 0$$

With these simplifications, the general mol balance equation simplifies to the following:

$$F_{j0} - F_j + r_j V = 0 \tag{5.10}$$

or

$$V = \frac{F_j - F_{j0}}{r_j} \tag{5.11}$$

We call this a *design equation*, because, with it, we can now do the first step in the design of the reactor—determine its volume, V—if we know the molar flow rate of j into and out of the reactor—and if we know r_j, which could be expressed by a constitutive relationship as a function of concentrations and temperature in the reactor (*anywhere in the reactor*, because, by our assumption of perfect mixing, there is no position dependence of any of the properties).

This steady-state backmix reactor is the simplest imaginable reactor: there is no dependence of the reaction rate on time and no dependence of the reaction rate on position. Alternatively, we could have a perfectly mixed reactor that was operated transiently, and we could have a steady-state mixed reactor that was not perfectly mixed. These would be different from the ideal reactor that we have just considered and would not be described correctly by Eq. (5.11).

The backmix reactor (whether operated transiently or at steady state) has other common names. It is also called an ideal *continuous stirred tank reactor* (abbreviated as CSTR, sometimes called "cee-star") (sometimes engineers speak loosely and refer to a mixed reactor as a CSTR even when it is not perfectly mixed; this is a mistake). A CSTR is also occasionally called a continuous-flow stirred tank reactor (CFSTR).

The next ideal reactors we consider can only be more complex than the steady-state backmix reactor; one of them will show a position dependence (with no time dependence), and the other will show a time dependence (with no position dependence).

The Perfectly Mixed Batch Reactor (A Reactor with Time Dependence but No Position Dependence)

Our second special case is a reactor with no flow in or out: a batch reactor, which we assume to be perfectly mixed. Thus, we make the following assumptions to simplify the general mol balance equation:

1. *No flow in* and *no flow out* (thus, by definition, a batch system).
2. *Perfect mixing* of the contents in the reactor.

Again, it will be convenient to define the volume of the reactor that contains reacting fluid as the system. Thus, the general mol balance equation that was written above,

$$F_{j_0} - F_j + \int^V r_j \, dV = \frac{dN_j}{dt} \tag{5.9}$$

simplifies to the following, because there is no flow into or out of the reactor:

$$\int^V r_j \, dV = \frac{dN_j}{dt} \tag{5.12}$$

We can make the following further simplification (as we did for the backmix reactor), because the mixing is assumed to be perfect and r_j is therefore a constant with respect to position (but not time): $\int^V r_j \, dV \rightarrow r_j V$. Consequently, the mol balance equation simplifies to the following:

$$V = \frac{1}{r_j} \frac{dN_j}{dt} \tag{5.13}$$

This is the design equation for the perfectly mixed batch reactor. The composition in the reactor changes continuously over time as reactants are converted into products, and this equation shows how. We stress that there is no position dependence in this case, but there is a time dependence.

The Steady-State Piston-Flow Reactor (A Reactor with Position Dependence but No Time Dependence)

Our third special case is an ideal flow reactor called a piston-flow (or plug-flow) reactor. We make the following assumptions:

1. *Steady state.*
2. *No mixing* in the reactor.

This model pertains to a reactor that is a tube or a pipe. Note that the assumption of no mixing is the opposite limiting case from the assumption of perfect mixing. This ideal reactor is represented schematically in Figure 5.4.

In the piston-flow reactor, the composition of the fluid changes continuously from inlet to outlet because reaction takes place. The farther downstream from the inlet a fluid element is, the more time it has had for reaction (i.e., for reactants to be converted into products).

To analyze the performance of this reactor, it is crucially important that we define the system effectively. It would not make sense to define the entire reactor as the system, because the composition of the fluid in the reactor changes with respect to position. Thus, the flow rate of j varies continuously from the inlet to the outlet (i.e., with distance down the pipe). We need to account for this variation. Consequently, we choose the system (control volume) to be a thin slice of the pipe volume oriented perpendicular to the axis, as depicted in Figure 5.5.

If the cross-sectional area of the pipe is A and A is a constant, and y is the distance down the pipe axis from the inlet, the thickness of the slice is Δy and the volume of the system $\Delta V = A\Delta y$.

Now we write the general mol balance equation over the control volume (using $\bar{r}_j$ to represent the average rate of reaction per unit volume in the volume element (the control volume), which is the cylindrical slice of the pipe). We need to use an average value of the rate in the slice because the composition of the reacting fluid changes over the distance between y and $y + \Delta y$, and hence the rate of reaction $\bar{r}_j$ changes over this distance:

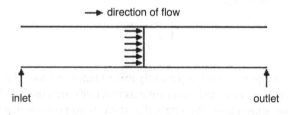

FIGURE 5.4 Schematic representation of a piston-flow reactor showing the flat velocity profile of the fluid; the horizontal arrows symbolize the velocities of various fluid elements, each one of which moves down the reactor parallel to the axis of the reactor and at the same rate.

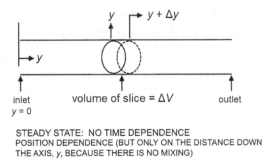

inlet   volume of slice = ΔV outlet
y = 0

STEADY STATE: NO TIME DEPENDENCE
POSITION DEPENDENCE (BUT ONLY ON THE DISTANCE DOWN
THE AXIS, y, BECAUSE THERE IS NO MIXING)

FIGURE 5.5   Schematic representation of the control volume (system) chosen for a piston-flow reactor with a circular cross section; the control volume is a slice of the reactor with a circular cross section perpendicular to the pipe axis.

$$F_j(y) - F_j(y + \Delta y) + \bar{r}_j A \Delta y = \left(\frac{dN_j}{dt} \right)_y \tag{5.14}$$

In this equation, the molar flow rate of j, F_j, is written as $F_j(y)$, and this notation means that F_j is evaluated at y (F_j is a function of y); correspondingly, $F_j(y + \Delta y)$ means F_j evaluated at $y + \Delta y$. Similarly, $\left(\dfrac{dN_j}{dt} \right)_y$ refers to the time rate of change of N_j evaluated at y.

Because we assume steady state, the right-hand (accumulation) term of Eq. (5.14) drops out:

$$F_j(y) - F_j(y + \Delta y) + \bar{r}_j A \Delta y = 0 \tag{5.15}$$

Now, following the usual procedures of calculus, we examine the limit as the slice becomes thinner and thinner: $\Delta y \to 0$. In this limit, we get the design equation for the steady-state piston-flow reactor.

$$\frac{dF_j}{dy} = r_j A \tag{5.16}$$

or

$$r_j = \frac{dF_j}{d(Ay)} = \frac{dF_j}{dV} \tag{5.17}$$

Notice that we have now written the equation for the point value of the rate, r_j. We no longer need an average because we have taken the limit as $\Delta y \to 0$. Remember that r_j is a function of y (and therefore a function of V).

UNDERSTANDING THE IDEAL REACTOR MODELS: SIMPLIFYING WITH THE ASSUMPTION OF NO VOLUME CHANGE ON REACTION

To better grasp the meanings of these models, we consider each of them with another simplification: we assume that there is no change in volume as a result of reaction. This constant-volume assumption is often very good when the reaction takes place in a liquid. Then we can write the following:

$$F_j = v C_j \tag{5.18}$$

where v is the (total) volume flow rate of the feed stream (which we called q in an earlier chapter) and C_j is the concentration of j, and

$$F_j = \text{molar flow rate of } j \left(F_j \ [=] \ \frac{\text{mol}}{\text{time}} \right) \tag{5.19}$$

(Also, remember the dimensions of $v \ [=] \dfrac{\text{volume}}{\text{time}}$ and the dimensions of $C_j \ [=] \dfrac{\text{mol}}{\text{volume}}$.)

Furthermore, with this assumption, we have $N_j = V C_j$, where N_j is the number of mols of j and V is the volume.

Now we simplify our three model equations to take account of the new assumption. We do this by substituting the above equations into the design equations for the three ideal reactors.

Case 1: Steady-State Backmix Reactor

$$V = \frac{F_j - F_{j0}}{r_j} = \frac{v\left(C_j - C_{j0}\right)}{r_j} \tag{5.20}$$

or

$$r_j = \frac{C_j - C_{j0}}{\dfrac{V}{v}} \tag{5.21}$$

We recognize that $\dfrac{V}{v} \left(\dfrac{\text{volume}}{\dfrac{\text{volume}}{\text{time}}} \right)$ is the *average residence time* of fluid in the reactor; let us call it $\bar{\tau}$, where the overbar indicates that it is an average value. Then,

$$r_j = \frac{C_j - C_{j0}}{\bar{\tau}} \left(\frac{\text{mol}}{\text{volume} \times \text{time}} \right) \tag{5.22}$$

Case 2: Perfectly Mixed Batch Reactor

Because the volume V is assumed to be constant,

$$r_j = \frac{1}{V}\frac{dN_j}{dt} = \frac{d\left(\dfrac{N_j}{V}\right)}{dt} \tag{5.23}$$

Thus, we see that

$$r_j = \frac{dC_j}{dt} \tag{5.24}$$

This is simply the time rate of change of the concentration of j, C_j. This equation gives us an extremely easy way to understand what the reaction rate per unit volume is in our special case of no volume change on reaction in a batch reactor—it is just the rate of change of concentration with respect to time.

Case 3: Steady-State Piston-Flow Reactor

We repeat the design equation from above and realize that, by our assumption of no volume change on reaction, the volume flow rate at steady state is constant:

$$r_j = \frac{dF_j}{dV} = \frac{d\left(vC_j\right)}{dV} = v\frac{dC_j}{dV} \; (v \text{ is a constant}) \tag{5.25}$$

$$r_j = \frac{dC_j}{d\left(\dfrac{V}{v}\right)} \tag{5.26}$$

V/v is just the residence time of a fluid element in the reactor. Because the flow is piston flow, each fluid element has the same residence time—it is in the reactor for the same time. We call this time τ (the overbar denoting an average is not needed because, as a consequence of the assumption of piston flow, all fluid elements have the same residence time τ). So,

$$r_j = \frac{dC_j}{d\tau} \tag{5.27}$$

We see that this matches the equation for the batch reactor, Eq. (5.24), with t (the clock time in the batch reactor) replaced by τ (the residence time in the piston-flow reactor). This is obviously correct, because the time that each fluid element has to react in the piston-flow reactor is just its residence time in the reactor, $\tau\left(= V/v\right)$. At a given volume flow rate v, the residence time τ increases in proportion to the reactor volume $V\left(= Ay\right)$.

Now it is helpful to compare the three design equations for the case of no volume change on reaction, as shown in Table 5.1. We need to understand what the mathematical statements mean.

Let us read these equations. Start with that for a batch reactor. Experimental results showing the composition of the reactor contents as a function of time (determined by chemical analysis) determine C_j as a function of t, as shown, for the example of a particular reaction at a particular temperature, in the sketch of Figure 5.6. In this case, the concentration of j increases continuously over time, and so j is obviously a product (the concentrations of reactants must decrease over time). Furthermore, it is evident that there was no product j present initially (because the curve fitting the data goes through $(0, 0)$), and we also see that the initial reaction mixture must have contained some reactants, because product j was formed. We see from the equation $r_j = \dfrac{dC_j}{dt}$ for this constant-volume batch reactor that we can evaluate the rate r_j at any time by drawing the tangent line to the curve fitting the data at that time (Figure 5.6) and taking its slope. In this case, the slope (r_j) decreases as t increases. This situation is typical; as t increases, the reactant is depleted (it is converted into product), and

TABLE 5.1

Summary of Model Equations for the Three Ideal Reactors (Assuming No Volume Change on Reaction)

Reactor Type	Model Equation (Design Equation)	Type of Equation	Assumptions	Limiting Cases
Backmix	$r_j = \dfrac{C_j - C_{j0}}{\left(\dfrac{V}{v}\right)}$	Algebraic	Flow, steady-state, perfect mixing	No dependence on position or time
Piston flow	$r_j = \dfrac{dC_j}{d\left(\dfrac{V}{v}\right)} = \dfrac{dC_j}{d\tau}$	Differential (calculus)	Flow, steady-state, no mixing (piston flow)	No dependence on time
Batch	$r_j = \dfrac{dC_j}{dt}$	Differential (calculus)	No flow (in or out), perfect mixing	No dependence on position

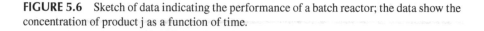

FIGURE 5.6 Sketch of data indicating the performance of a batch reactor; the data show the concentration of product j as a function of time.

because the rate typically decreases as the reactant concentration decreases (e.g., one possibility is $r_j = kC_i$, where i is a reactant—this is a constitutive relationship), the reaction slows down as the reaction proceeds in the batch reactor.

Recalling that the reaction rate r_j depends on concentrations of the species present in the reactor (and on temperature—which, however, was assumed to be constant in this example), we see that the slopes of the curve at various points give us information about how the rate depends on the concentrations of these species; we would have to know the initial concentration of each species and the stoichiometry of the reaction to determine the concentration of each species at each time. Thus, data such as those represented in Figure 5.6 provide information about the constitutive relationship, that is, how r_j depends on the concentrations of the species in the reactor. If we were to change the concentrations of the reactants and products present initially, we could determine more about the constitutive relationship, that is, how r_j depends on the concentrations, and if we were to change the reactor temperature systematically, we could determine still more about the constitutive relationship, that is, how r_j depends on the temperature.

Now, consider the piston-flow reactor. In Figure 5.7, we see the same curve as in Figure 5.6. The difference is that in the piston-flow reactor, the concentration C_j depends on the position in the reactor (y; or, more precisely in this figure, [A/v] times y), whereas, in the perfectly mixed batch reactor, C_j depends on time. Thus, in the one case, the fundamental independent variable is position, and in the other case, it is time.

More importantly, mathematically, the equations describing these two reactors are identical in form—the variable t in the batch reactor equation is replaced by τ in the piston-flow reactor equation. The elapsed time t in the batch reactor is what we would measure with a clock. The residence time τ in the piston-flow reactor is just how much time each fluid element spends in the reactor—it increases with length (volume) of the reactor (other things being equal) and decreases with increasing volume flow rate of the reactant stream (other things being equal). We would measure τ in an experiment as the volume of the reactor divided by the volume flow rate of the stream flowing through it.

Again, the rate r_j is the slope of the curve showing C_j vs. τ (Figure 5.7). At any position in the reactor (at any V or any y), we find the rate from the slope of the tangent line. In this case, the rate decreases from inlet toward outlet as the concentration of reactant i decreases from inlet to outlet.

The relationship between C_i and C_j is determined by the stoichiometry of the reaction. For example, the simplest stoichiometry is

$$A \rightarrow B \qquad (5.28)$$

such as in the isomerization reaction

$$n\text{-propyl alcohol} \rightarrow \text{isopropyl alcohol} \qquad (5.29)$$

Here, there is only one reactant; i is A, and there is only one product, j is B. Thus, the dependence of the concentration of the reactant i (i.e., A) on τ is as shown in Figure 5.8. Because of the simple one-to-one stoichiometry, the curves in Figures 5.7 and 5.8 are mirror images of each other.

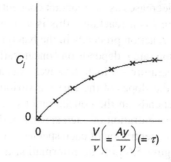

FIGURE 5.7   Sketch of data indicating the performance of a piston-flow reactor; at steady state, the concentration of product j depends on the position in the reactor (y, or, more precisely, A/ν times y).

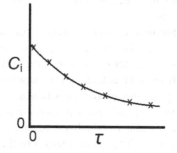

FIGURE 5.8   Sketch of data indicating the performance of a piston-flow reactor; at steady state, the concentration of reactant i depends on the residence time in the reactor, τ, and the position in the reactor (y, the distance from the inlet, or, more precisely, A/ν times y).

UNDERSTANDING BACKMIX AND PISTON-FLOW REACTORS

Notice from Table 5.1 that the equations describing the batch and piston-flow reactors are differential equations (calculus) and that the equation describing the backmix reactor is algebraic. The equation for the backmix reactor does not represent a continuous change in concentration C_j with respect to either (a) time or (b) position, because we assumed (a) steady state and (b) perfect mixing.

How do we understand this behavior? Because the mixing is perfect, the composition (and C_j) must be the same at every position in the reactor, and one of those positions is at the outlet stream. To repeat, the value of C_j at the outlet must be the same as the value of C_j everywhere in the reactor (and we could make the same statement about the concentration of each species in the reactor). When reaction occurs, the value of C_j in the reactor must be different from the value of C_j in the feed. And so, according to our model, there is a STEP CHANGE in concentration C_j *right at the inlet*; the feed stream has one composition (and the concentration of $j = C_{j0}$), and the reactor contents (everywhere) have another composition (and the concentration of $j = C_j$).

The concept of a step change in this context requires some reflection. What it means is that the feed stream, just as it enters the reactor, is instantaneously mixed with the fluid in the reactor, so that all the change in composition occurs right at the position where the feed enters the reactor. This is of course a simplification; there must be some region in the fluid where there is a variation in composition with respect to position. But the assumption of a step change in concentration is often very good, and the region where the composition is not uniform is often too small to measure easily.

Notice that the step change in concentration corresponds to what we might call a difference equation (the design equation for the steady-state backmix reactor)—that is, an algebraic equation in which a change in concentration appears—whereas the gradual change in concentration with respect to position in the piston-flow reactor corresponds to a differential equation (calculus). Recognize that the two design equations for the two flow reactors correspond to the only logical limiting cases of mixing—perfect mixing for the backmix reactor and no mixing for the piston-flow reactor.

The assumption of no mixing in the piston-flow reactor means that each fluid element remains in this reactor for the same length of time. In other words, there is only one, unique, residence time. If a tracer (say, a dye) is injected as a pulse into a piston-flow reactor at the inlet at time equal to zero, it will appear as a pulse at the exit of the reactor after a time equal to the residence time of any fluid element in the reactor, that is, after a time equal to the reactor volume divided by the volume flow rate, V/v.

In contrast, in the backmix reactor, because it is perfectly mixed, every fluid element has an equal probability of being in the exit stream—independent of whether it just entered the reactor or has resided there for a long time. Thus, in the backmix reactor, there is a distribution of residence times, and not just one. We refer to the average residence time in the backmix reactor, which we call $\bar{\tau}$; remember that the overbar indicates the average. The value $\bar{\tau}$ is $\dfrac{V}{v}$, the volume of the reactor divided by the volume flow rate. Can you figure out, by our assumption, what is the shortest residence time of a fluid element in a backmix reactor?

Let us examine these concepts a bit more deeply. Consider a backmix reactor that has been running at a steady state with a feed consisting of a solvent with a low concentration of an unreactive component (call it a tracer) dissolved in it. Assume that at time = zero, the feed stream is switched to pure solvent, without the tracer, and that the volume flow rate remains unchanged. After that time, no tracer enters the reactor, and, because the flow continues at an unchanged volume flow rate (we assume that the tracer concentration is so low that it negligibly affects the solution density), the tracer in the reactor is purged out (washed out). The initial concentration of tracer in the exit stream is the initial value, that is, the value that was attained prior to stopping the tracer flow in the inlet. The concentration of tracer in the outlet declines with time as it is washed out of the reactor.

If the mixing of the reactor contents is perfect, we can predict the concentration of tracer in the reactor (and therefore in the exit stream) as a function of time, just by using the general mol balance Eq. (5.9) for the non-steady-state backmix reactor for the special case of no flowing reactant and no reaction. Consequently, we simplify the equation for species j (now the tracer) by setting the terms *except F_j* and dN_j/dt equal to zero. For simplicity, let us again assume that $N_j = VC_j$ and $F_j = vC_j$. Then, Eq. (5.9) simplifies to

$$\frac{d(VC_j)}{dt} = -vC_j \tag{5.30}$$

We integrate from time $t = 0$ when C_j is equal to its initial (steady-state) value (which we call C_{j0}), getting a result showing how the tracer concentration decays with time:

$$C_j = C_{j0}e^{-(V/v)t} \tag{5.31}$$

(Can you check the calculus here?) Notice that we have now used the subscript 0 to designate an initial (rather than an inlet) value. We also do this for analysis of a batch reactor. This result tells us that a plot of C_j on a logarithmic scale as a function of time after the change in feed composition will be a straight line, provided that the assumptions are valid. Remember that one of the assumptions is that the mixing is perfect. If all the assumptions except for this one are correct, and if we see that if the data do fall on a straight line, then we conclude that the mixing is perfect. Alternatively, if all the assumptions except for this one are correct, and if we see that the data do not fall on a straight line, then we conclude that the mixing is imperfect.

Thus, we have a basis for evaluating the assumption of perfect mixing in a reactor: to test for perfect mixing, we start with the reactor filled with a solvent containing a low concentration of tracer and start a steady-state flow of the solvent without tracer into the reactor with a steady-state flow of product at the same volume flow rate out of the reactor, and then we measure the concentration of the tracer in the exit stream as a function of time, and make the semilogarithmic plot and check whether it is linear.

Determining Concentrations of Reactants and Products in Reactors

Reaction rates depend on concentrations of reactants, products, and perhaps other components in reactors, such as catalysts, promoters, and components that hinder reactions, such as inhibitors and poisons (we explain below what these are). To analyze the performance of reactors (and therefore to design reactors), we therefore need to measure concentrations of the reactants, products, etc. There are many analytical methods used by chemists to make these measurements. For example, samples can be drawn from reactors and analyzed off-line by techniques such as gas chromatography and liquid chromatography. Physical properties such as electrical conductivity that are correlated with concentrations of various components can be measured continuously with probes in the reactors; these need to be calibrated with standards. Spectra of reacting mixtures can similarly be measured continuously; these also need to be calibrated, and sometimes the calibrations are simple, so that the intensity of a band at a particular wavelength in a spectrum (e.g., the area under an ultraviolet or visible or infrared band) may be proportional to the concentration of a particular species. It is common for experimentalists to record physical properties of solutions that provide real-time determinations of concentrations of compounds in the solutions and to use the data to determine how rates of reactions depend on those concentrations. This topic is developed further below.

Example 5.2 Use of Data from a Backmix Reactor to Predict the Performance of a Batch Reactor

Problem statement: The reaction $A \rightarrow B$ took place at a constant temperature and with no volume change in a steady-state backmix reactor with a volume of 10,000 L. The feed to the reactor was pure A with a concentration of 10.0 mol/L. As shown in the table below, various concentrations of B in the product stream were observed continuously at various volume flow rates of the feed. Use the results to find the time required for a change in the concentration of A in a perfectly mixed batch reactor from 10.0 to 1.0 mol/L if the batch reactor is operated at the same temperature as the backmix reactor.

C_B (mol/L)	$10^4 \times r_B$ (mol/(L × h))	$10^{-4}/r_B$ ((L × h)/mol)
9.00	0.99	1.01
8.00	1.41	0.71
6.00	2.00	0.50
4.00	2.44	0.41
2.00	2.80	0.36
1.00	3.00	0.33

Solution

We are given data for the steady-state CSTR, including the volume V, the volume flow rate v, and the concentration of B in the product, C_B, sufficient to find the rate of reaction in the CSTR under each set of operating conditions. We find the rate, r_B, by applying the corresponding design equation. Then, with these values of r_B, we can use the design equation for the batch reactor to find the time required for the stated conversion. According to the CSTR design equation, $r_B = (C_B-C_{B0})/(V/v)$; thus, values of r_B were calculated and are included in the table above.

To find the time required for the stated conversion in the batch reactor, we need the appropriate design equation, which is the following:

$$\int dt = t = \int \frac{dC_B}{r_B}$$

Thus, we need to evaluate the integral by using the data in the table showing r_B as a function of C_B; we do this graphically by plotting $1/r_B$ (values are calculated and shown in the table) vs. C_B (as shown in the figure below) and finding the area under the curve; the limits of the integration are from the initial value of C_B (zero) to the final value, 9.0 mol/L, found from the stoichiometry of the reaction (when $C_A = 1.0$ mol/L, $C_B = 9.0$ mol/L). The plot is shown below; the area under the curve is approximately 4.3×10^4h; this is the required time.

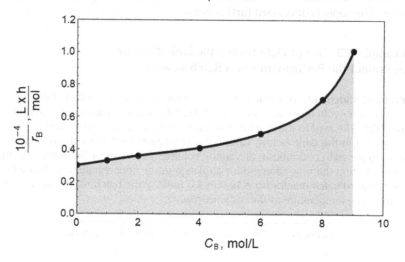

Example 5.3 Analysis of Batch Reactor Data

Problem statement: Consider the reaction A $\rightarrow$ B taking place at a constant temperature and with no volume change in a perfectly mixed batch reactor initially containing only A and inert solvent. The data determine the concentration of A, C_A, as a function of time t, as shown in the following table:

C_A (mol/L)	t (s)
1.08	0
0.53	20
0.26	40
0.13	60
0.066	80
0.032	100
0.015	120

A. Plot C_A and C_B as a function of t and explain the shapes of the plots.

Solution of Part A

Because the reaction stoichiometry is simply A $\rightarrow$ B, the rate of disappearance of A must equal the rate of appearance of B: $-r_A = r_B$. We reason, as above, from the mol balance equation and recognize the simplifications that follow because there is no flow into and no flow out of the reactor, and because the mixing is perfect:

$$F_{A0} - F_A + \int r_A V = \frac{dN_A}{dt}$$

We simplify as in the text above to determine the following equations using the statement that

$$-r_A = r_B:$$

$$r_A = \frac{1}{V} \times \frac{dN_A}{dt} = \frac{d\frac{N_A}{V}}{dt} = \frac{dC_A}{dt}$$

$$\frac{dC_B}{dt} = r_B = -r_A = -\frac{dC_A}{dt}$$

We know from the stoichiometry that

$$C_B = C_{B0} + (C_{A0} - C_A) = (C_{A0} - C_A)$$

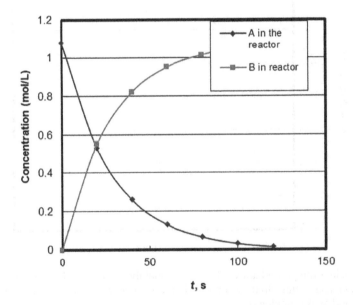

where the subscripts 0 refer to initial values. The plot of the data from the table is shown above. We see that the two curves are mirror images of each other, corresponding to the stoichiometry, and that the curves bend and approach horizontal lines at long times. As the slopes determine the reaction rate, we thus see that the rate approaches zero at long times.

B. Estimate $-dC_A/dt$ at various values of t and make a graph showing these values as a function of C_A. Explain the shape of the graph qualitatively.

Solution of Part B

From above, the slope of the tangent line of the plot of C_A vs. t is equal to the rate, $-dC_A/dt$. We use a graphical method to determine the following values of the slope:

t, s	dC_A/dt, mol/(L × s)
0	−0.0372
20	−0.0204
40	−0.0092
60	−0.0043
80	−0.0022
100	−0.001
120	−0.0003

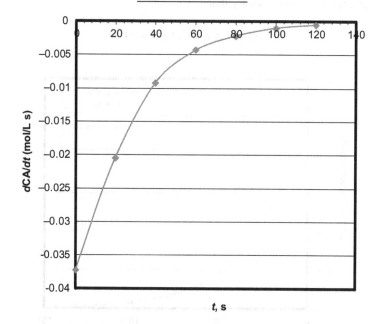

These values are plotted above. They show that the value of $-dC_A/dt$ decreases as time increases—thus, that the rate of disappearance of A decreases as time proceeds and as A is depleted.

C. Determine an empirical equation for dC_A/dt as a function of t.

Solution of Part C

We proceed empirically and iteratively, first assuming a simple dependence of dC_A/dt on t and testing how well the data conform to the assumption, and then proceeding to another assumption and so on. First we assume that dC_A/dt is independent of t; just by examining the plot, we realize that this is not a good approximation, as dC_A/dt is not constant; obviously, dC_A/dt depends on C_A. So let us assume that dC_A/dt is directly proportional to C_A:

$$\frac{dC_A}{dt} = -r_A = k \times C_A$$

We know from integrating this equation that a plot of C_A on a logarithmic scale vs. t will be a straight line if the data agree with this assumption. And so we make this plot as shown below:

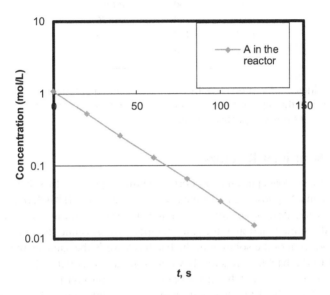

t, s

The data fall close to a straight line, showing that the assumption is good. The slope of the line determines $k-0.0354\,s^{-1}$. Thus,

$$\frac{dC_A}{dt} = -0.0354 C_A$$

By integrating this equation, we find $C_A = C_{A0} \times \exp(-0.0354t)$ where $C_{A0}=1.08$ mol/L.

D. As a check, use the integrated equation of part C to find an empirical equation for C_A as a function of t and compare it with the data.

Solution of Part D

The integrated equation is

$$C_A = C_{A0} \times \exp(-0.0354t)$$

The comparison with the data is as shown in the table below:

C_A (mol/L)	t (s)	Estimated Value of C_A (mol/L)
1.08	0	1.08
0.53	20	0.532
0.26	40	0.262
0.13	60	0.129
0.066	80	0.0636
0.032	100	0.0313
0.015	120	0.0154

The values of C_A predicted by the empirical equation are close to those observed, but the agreement is not perfect because there are errors in the data and the equation is an approximation.

DESIGN OF SOME IDEAL REACTORS

Let us use these concepts and the corresponding equations to solve some reactor design problems. Suppose we make measurements in a (small) batch reactor in a laboratory and wish to design (i.e., find the volume of) a (large) commercial-scale backmix reactor for the same reaction at the same temperature. Assume for simplicity that the initial composition of the solution in the batch reactor is the same as the composition of the feed to the backmix reactor. If we have made measurements to determine C_j as a function of t in the batch reactor operated at a constant temperature—with the values of C_j determined by chemical analysis of samples taken from the reactor at various times, t, then we can plot C_j vs. t, as before (Figure 5.9).

Now, suppose we want to find the volume of a backmix reactor with a feed volume flow rate of v and the value of C_j in the product equal to the value $(C_j)_{prod}$ shown in Figure 5.9. To design the backmix reactor, we calculate the value of r_j as the slope of the above curve at $C_j = (C_j)_{prod}$, plug it into the model equation for the backmix reactor with the values of C_j ($= (C_j)_{prod}$) and C_{j0} (presumed to be given) and the value of v (presumed to be given) and solve for the reactor volume V.

An essential point in this reactor design is that if the feed to the flow reactor matches the initial composition of the batch reactor, then the composition of the

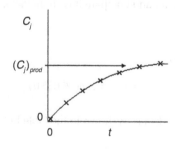

FIGURE 5.9   Sketch of data indicating the performance of a constant-temperature perfectly mixed batch reactor; after a particular time, the product concentration reaches a particular value, as indicated, for example, by the symbol x to the right of the arrowhead.

fluid in the batch reactor at the time corresponding to the value of $C_j = (C_j)_{prod}$ is the same as the composition in the backmix reactor having the same value of C_j. This statement is true because the change in composition of the fluid in either reactor is determined entirely by the stoichiometry of the reaction, which is of course not dependent on the reactor—just as the rate r_j is not dependent on the reactor, but only on the concentrations of the species involved in the reaction and the temperature.

Think of the point restated this way: *the molecules don't know what kind of reactor they are in.*

We can generalize strongly, with the preceding example providing just one of many possible illustrations of the point: *data from any one type of ideal reactor can be used to design any (other) ideal reactor.* To do the design, we use the model equation for each type of reactor and data from the type of reactor that has been used to generate the data; in other words, *we use the model equation for the reactor that was used to generate the data to analyze the data, and then we use the model equation for the target reactor (the reactor to be designed) with the results of the analysis to design the target reactor.* What links the equations for the two ideal reactors is the reaction rate r_j. This term appears in each design equation (that for each kind of ideal reactor). We can determine values of r_j from data obtained with any kind of ideal reactor and use the appropriate values of r_j to find the volume of any other kind of ideal reactor. To do the calculations, we use calculus for the batch and piston-flow reactors and algebra for the backmix reactor.

DETERMINING CONSTITUTIVE RELATIONSHIPS FROM IDEAL REACTOR DATA

We can use data from any of the three ideal reactor types to determine the constitutive relationship for any reaction carried out in that ideal reactor. For example, for the reaction $A \rightarrow B$, taking place at a constant temperature with no volume change on reaction, consider data showing the concentration of a reactant A in a constant-temperature batch reactor as a function of the time t, as sketched in Figure 5.10.

The reaction stoichiometry is $A \rightarrow B$; i is A, and j is B. Therefore, for every mol of A converted, one mol of B is formed. Therefore, $-\dfrac{dC_A}{dt} = \dfrac{dC_B}{dt}$.

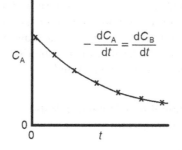

FIGURE 5.10 Sketch of conversion of A to B in a batch reactor at a constant temperature with no volume change on reaction. The slope of the curve at any particular value of the time t determines the rate of the reaction corresponding to that particular concentration of A and the corresponding concentration of B.

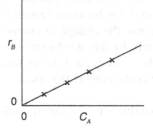

FIGURE 5.11 Sketch showing determination of a constitutive relationship for the reaction A → B.

Our goal now is to find a constitutive relationship for this reaction. We know that reaction rates depend on temperature and concentrations of species involved in the reaction. The data of Figure 5.10 are for a reaction at a constant temperature, and so these data tell us only about how the rate depends on the concentration of A. We assume now, for simplicity, although it is not always true, that the rate of the reaction depends on the concentration of the reactant A and not on the concentration of the product B.

To find an equation for this dependence, we take the slopes of the curve at various values of C_A. These are values of $\dfrac{dC_A}{dt}$ (all negative); because $-\dfrac{dC_A}{dt} = \dfrac{dC_B}{dt}$, the rate of formation of B is positive and the rate of formation of A is negative (but the rate of disappearance of A is positive). We plot $r_B (= r_j) = \dfrac{dC_B}{dt}\left(\dfrac{dC_j}{dt}\right)$ as a function of C_A, as shown in the sketch of Figure 5.11.

Then we find an equation to fit the data using methods illustrated earlier in this book. For example, the sketch of Figure 5.11 implies $r_B = kC_A$ (where k is a constant) corresponding to what is called a first-order reaction. We say that the reaction is first-order in A. This equation is an example of a *reaction rate equation*.

Constitutive equations such as this are useful in design. We would proceed to use the equation by combining it with a reactor design equation (model equation). Examples follow.

We stress that the constitutive relationship for a reaction with the one-to-one stoichiometry A → B is not always $r_B = kC_A$. The form of the constitutive relationship cannot be determined from the reaction stoichiometry—it does not depend on the stoichiometry. It depends on the chemistry—what A and B are—and the conditions of the reaction. We can find the form of the constitutive relationship only by doing experiments for the particular compounds A and B.

Example 5.4 Determination of Reaction Rates from Reactor Performance Data

Problem statement: The results shown in the following table were obtained for the reaction A → B taking place with no volume change in a perfectly mixed batch reactor at a constant temperature. The rate of the reaction depends on the concentration of A but not on the concentration of B. The initial composition was

A with a concentration of 1.00 mol/L in an inert solvent. Because the stoichiometry of the reaction indicates that one molecule of A reacts to give one molecule of B, we can write that the rate of disappearance of A, $-r_A$, equals the rate of appearance of B, r_B.

Time (h)	Concentration of A in Reactor (mol/L)	Concentration of B in Reactor (mol/L)
0	1.00	0.00
25	0.77	0.23
50	0.60	0.40
75	0.47	0.53
100	0.37	0.63
125	0.29	0.71
150	0.23	0.77
175	0.18	0.82

Problem A. Estimate the rate of reaction r_B in the batch reactor when the product contains 0.37 mol/L of A.

Solution

The plot of concentration vs. time is shown as follows:

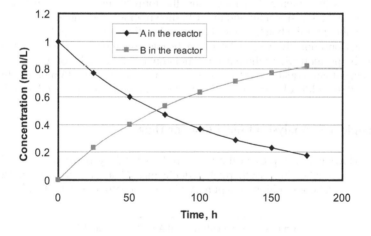

Because the reaction is A → B, we realize that $-r_A = r_B$.

As in Example 5.2, we recognize that $r_A = \dfrac{1}{V} \times \dfrac{dN_A}{dt} = \dfrac{dC_A}{dt}$, and therefore the slope of the tangent line of the plot of C_A vs. t shown above is equal to r_A at any particular time.

When the product contains 0.37 mol/L of A, the slope of this tangent line is -0.0036 mol/(L × h), and so $r_A = -0.0036$ mol/(L × h) and $r_B = 0.0036$ mol/(L × h).

Problem B. Estimate the rate of reaction r_B at a position in a steady-state piston-flow reactor operated at the same temperature as the batch reactor when the feed to the reactor is A with a concentration of 1.00 mol/L in the inert solvent and the reaction solution at that position contains 0.37 mol/L of A.

Solution

For a steady-state piston-flow reactor, the general mol balance equation:

$$F_A(y) - F_A(y + \Delta y) + \bar{r}_A \times A \times \Delta y = \left(\frac{dN_A}{dt} \right)_y$$

can be simplified as follows:
For steady state,

$$\left(\frac{dN_A}{dt} \right) = 0$$

and in the limit as $\Delta y \to 0$,

$$r_A = \frac{dF}{dV} = \frac{dC_A}{d\left(\dfrac{V}{v} \right)} = \frac{dC_A}{d\tau}$$

where v is the volume flow rate. Thus, the form of this equation matches that describing the batch reactor and the data shown above; the only difference is that the variable t for the batch reactor is replaced by the variable τ for the piston-flow reactor. Consequently, we can determine the rate from the C_A vs. t plot to determine r_A; at the position where the reactant fluid contains 0.37 mol/L of A, the slope of the tangent line equals -0.0036 mol/(L × h), and so $r_A = -0.0036$ mol/(L × h) and $r_B = 0.0036$ mol/(L × h), as expected from the result of the first part of this problem.

Example 5.5 Analysis of Ideal Reactor Data

The following data represent the reaction A → B taking place with no volume change on reaction in an isothermal, perfectly mixed batch reactor. The feed to the reactor (initial composition) was pure A with a concentration of 10.0 mol/L.

Time (s)	Concentration of A in Reactor (mol/L)
0	10.0
50	7.7
100	6.0
150	4.7
200	3.7
250	2.9
300	2.3
350	1.8

Problem statement A: Suppose that the volume of the reactor is doubled and that the initial composition and all the other conditions are kept the same. Estimate the rate of reaction r_B in the reactor when the concentration of A is 3.7 mol/L.

Solution

Because there is no volume change on reaction and temperature is constant, the rate of reaction $-r_A = dC_A/dt$. Because rate depends on concentration and temperature only, doubling the reactor volume does not change the rate in a batch reactor. So we get the rate as slope of C_A vs. t curve at $C_A = 3.7$ mol/L.

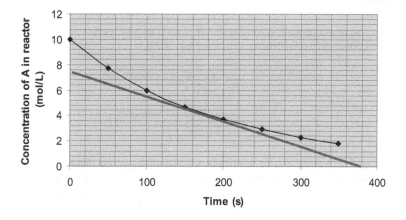

Slope of straight (tangent) line $= -(7.5-0)/(380-0) = -0.020$ mol/(L × s). Consequently,

$$-r_A = r_B = 0.020 \text{ mol/(L × s)}.$$

Problem statement B: Estimate the rate of reaction r_B in a steady-state CSTR when the feed is pure A, the temperature is the same as that in the batch reactor, and the product contains 3.7 mol/L of A.

Solution

Because the rate depends only on concentration and temperature and because the concentration of any component in the product of the CSTR equals the concentration everywhere in the reactor—and the concentration of A is given to be 3.7 mol/L, we see that the rate is the same as that just determined for the batch reactor at this concentration of A, namely, 0.020 mol/(L × s).

EXAMPLES OF REACTOR DESIGN CALCULATIONS
USING A CONSTITUTIVE RELATIONSHIP

The following example problems illustrate a common goal in engineering: to take data from one system (often one in a laboratory, hence, small and convenient and safe to use) and use them to design another system (often large, and intended to meet goals of safe, profitable production).

Example 5.6 Further Analysis of Ideal Reactor Data

Problem statement A: Given a reaction A → B at a constant temperature taking place with no volume change on reaction and characterized by the constitutive relationship $r_B = kC_A$, with $k = 0.10\,min^{-1}$, find the average residence time for the reaction in a steady-state backmix reactor operated at the same temperature as that used to determine the constitutive relationship, for a conversion of C_A of 90%, with a concentration of A in the feed stream, C_{A0}, of 10 mol/L and a concentration of B in the feed stream, C_{B0}, of 0. If the reactor volume is 1,000 L, find the volume flow rate of the feed stream.

Solution

We use the design equation for the steady-state backmix reactor (from Table 5.1) and the constitutive relationship, repeating Eq. (5.21)

$$r_j = \frac{C_j - C_{j0}}{\left(\dfrac{V}{v}\right)} \tag{5.21}$$

$$r_B = 0.10C_A \tag{5.32}$$

We recognize that for the backmix reactor, the composition everywhere is the same and equal to that of the exit (product) stream. Therefore, because 90% of the reactant A is converted, the change in concentration of A in the reactor is $(10.0 - 1.0) = 9.0$ mol/L. Thus, using Eq. (5.32), we calculate the rate $r_B = 0.10C_A = 0.10 \times 1 = 0.10$ mol/(L × min). Using these values in Eq. (5.21), we find that $\left(\dfrac{V}{v}\right) = 9/0.10 = 90\,min$. Given that $V = 1,000$ L, we calculate the volume flow rate v as $1,000/90 = 11.1$ L/min.

Problem statement B: Given this same reaction with the same constitutive relationship, find the time for reaction in a perfectly mixed batch reactor at the same temperature to give a conversion of 99.9%, assuming that the initial reactant solution is A at a concentration C_{A0} of 10 mol/L with no B present.

Solution

From the development above, we have

$$-r_A = r_B = \frac{dC_B}{dt} = -\frac{dC_A}{dt} \tag{5.33}$$

Rearranging this, we have

$$dt = \frac{dC_B}{r_B} = \frac{dC_B}{kC_A} \tag{5.34}$$

Consequently,

$$dt = -\frac{dC_A}{kC_A} \tag{5.35}$$

Now, to find the time t, we integrate both sides of the equation (recognizing that k is a constant because the temperature is a constant):

$$\int_0^t dt = t = -\frac{1}{k} \int_{C_{A0}}^{C_A} \frac{dC_A}{C_A} \tag{5.36}$$

or

$$t = \frac{1}{k} \ln \frac{C_{A0}}{C_A} \tag{5.37}$$

We can thus calculate the time t by realizing that when 99.9% of the reactant A is converted, $\frac{C_{A0}}{C_A} = 1000$. Thus, using the value of k determined above, we find $t = (1/0.10)\ln(1000) = 69.0\,\text{min}$.

Problem statement C: Given this same reaction with the same constitutive relationship, find the residence time for reaction in a steady-state piston-flow reactor at the same temperature to give a conversion of 99.9%, assuming that the composition of the feed to the reactor is A at a concentration C_{A0} of 10 mol/L, with no B.

Solution

We see from Table 5.1 that the equation for the piston-flow reactor is the same as that for the batch reactor, except that the residence time appears instead of the clock time. Thus, by comparison with the preceding example, we see that the residence time in the piston-flow reactor is the same as the time required for reaction in the batch reactor, 69.0. min. What this means is that each fluid element in the piston-flow reactor acts like its own, independent batch reactor—with reaction occurring for the length of time the fluid element resides in the reactor.

DESIGN OF REACTORS FOR DETERMINING KINETICS OF EXTREMELY FAST REACTIONS

Rates of chemical reactions span many orders of magnitude, with some of the slowest taking place in geology, with time frames of millennia (e.g., the time for reaction to proceed, for example, to 50% completion). Other reactions take place on time scales orders of magnitude less than a second. When reactions are extremely fast, measurement of rates and rate equations is challenging because significant change may take place before measurements of concentrations can be made. In the next paragraph, we consider a design that facilitates determination of rates of reactions taking place on time scales of milliseconds.

Some spectrometers allow recording of quantitative data that determine concentrations that change within time frames of milliseconds. Many reactions, including biological reactions, take place on these time scales. Conventional batch reactors are impractical because substantial reaction takes place before the reactants are introduced into the reactor and before they can be mixed thoroughly. Similarly, conventional flow reactors are impractical because reaction takes place in tubes upstream of the reactor. A design that overcomes these limitations is called a stopped-flow reactor. Consider an example in which two compounds react with each other. According to the stopped-flow reactor design, each reactant in solution flows in a separate tube, and these flow together at steady state into a reactor which is a spectroscopy cell where they are rapidly and continuously mixed—with the mixing being perfect. As soon as the two streams reach this reactor, reaction starts. The flow system is brought to a steady state, as shown by the lack of time dependence of the spectrum of the solution in the spectrometer cell that is the reactor, and then the flow into and out of the cell that is the reactor is abruptly stopped. The equipment must allow this change to take place on a time scale that is short by comparison with the time characteristic of the chemical reaction. Once the valves are closed, the cell becomes a batch reactor, and if it is perfectly mixed and the spectrometer records data quickly enough, the spectra determine concentrations of reactants and/or products in a period of a few milliseconds. Thus, a flow system is used to attain a steady state, and when the valves stop the flow in and out of the cell/reactor, the cell/reactor that was part of the flow system becomes a batch reactor, and concentration vs. time data can be collected to determine conversion vs. time data. Efficient mixing in the cell/reactor is essential to allow straightforward determination of such data, and practical designs allow this by design of the fluid flow pattern.

Generalizations About Constitutive Relationships for Reactions

Remember that constitutive relationships are determined by experiment; they are not general, but instead pertain to particular situations. In the case of chemical reactions, each reaction is characterized by its own constitutive relationship, called a reaction rate equation (or a representation of the *kinetics* of the reaction). Rate equations may change in form as the temperature changes or as the concentrations change. We have seen examples above of simple rate equations. Many reactions are characterized by simple rate equations.

Many experiments done for many reactions lead to some generalizations about the forms of rate equations, as summarized below.

It is usually true that

$$r_i = k \prod_j C_j^{\alpha_j} \tag{5.38}$$

where k is a temperature-dependent (but not concentration-dependent) term called a reaction rate constant (or simply a rate constant) and $\prod_j C_j^{\alpha_j}$ is a concentration-dependent (but not temperature-dependent) term in which the exponents on the concentrations are often integers (typically, 0, 1, or 2), or half-integers (typically, $\frac{1}{2}$ or $3/2$).

The exponents α_j are called reaction orders. For the reaction A → B, if r_B = rate of formation of B = kC_A, as in an example above, then the order of reaction in A is 1; the order of reaction in B is 0; and the overall reaction order is 1. If $r_B = kC_A^2$, then the order of reaction in A is 2; and the order of reaction in B is zero.

The term k is called a rate constant, but it is not really a constant. It depends on temperature (but not on concentration). The units of k depend on the reaction order.

Thus, these units might be s^{-1} in for a first-order reaction and $\dfrac{L}{mol \times s}$ for a second-order reaction.

The results of many experiments show that rate constants k are exponentially dependent on temperature:

$$k = Ae^{-E/RT} \tag{5.39}$$

where A is the pre-exponential factor, having dimensions of k. In this equation, called the Arrhenius equation, E is the activation energy, typically expressed in units of $\dfrac{kJ}{mol}$ or $\dfrac{kcal}{mol}$, R is the gas constant (as in the ideal gas law), and T is the absolute temperature (in units of K or R). This equation tells us that a plot of k on a logarithmic scale vs. $1/T$ is a straight line with a slope of $-E/R$.

Keep in mind that rate equations are specific to particular reactions and must be determined experimentally. But in some simple cases, we can anticipate their forms. We consider below one of these simple cases to show how the form of the rate equation makes sense.

Usually, the higher the concentration of a particular kind of molecule, the faster it reacts (but this is not always true). Consider the case in which two molecules react with each other as a result of a collision between them in the gas phase. Then, the rate of the reaction is proportional to the concentration of each. Why is this plausible? Because, to react with each other, the two molecules, A and B, must come in contact with each other, they must collide with each other to facilitate the making and/or breaking of chemical bonds that constitute the reaction. The frequency of collisions involving the two kinds of molecule is proportional to the concentration of each, at least when the molecules are present in an ideal gas. This is a result shown by the fundamental chemistry—the kinetic theory of gases. So, the collision frequency is proportional to the number of molecules of A (or mols of A) per unit volume times the number of molecules of B (or mols of B) per unit volume, and often the rate of the reaction of A with B is proportional to the collision frequency, giving a second-order reaction (but this statement does not mean that every collision of A with B is successful in leading to a reaction—instead, the molecules sometimes just bounce off each other).

Matters are not always as simple as this, because often reactions take place through sequences of steps, whereby, for example, A reacts with B via some intermediate species, different from either A or B.

Example 5.7 Analyzing Data to Determine an Activation Energy

Problem statement: Consider the following reaction (a saponification reaction—can you figure out why this is a good name?) that was carried out in an aqueous solution in an isothermal batch reactor at various temperatures. The data were obtained by K. Das, P. Sahoo, M. Sai Baba, N. Murali, and P. Swaminathan (Kinetic studies on saponification of ethyl acetate using an innovative conductivity-monitoring instrument with a pulsating sensor, *Int. J. Chem. Kinet.*, **2011**, *43*, 648–656):

$$CH_3COOC_2H_5 + NaOH \rightarrow CH_3COONa + C_2H_5OH \qquad (5.40)$$

The people who made the measurements determined values of the rate constant of the reaction as a function of temperature, as summarized in the following table:

Temperature (K)	Rate Constant k (L (mol × s))$^{-1}$)
303.0	0.162
311.7	0.266
313.3	0.280
317.5	0.346
323.0	0.437
328.5	0.603

Show that these data are consistent with the Arrhenius equation. Determine the activation energy of the reaction.

Solution

To check for the consistency with the Arrhenius equation, we realize from Eq. (5.39) that we should make a plot of the rate constant on a logarithmic scale vs. the inverse of the absolute temperature, expecting that if the data are consistent with the Arrhenius equation, the data will fall near a straight line. Here is the plot:

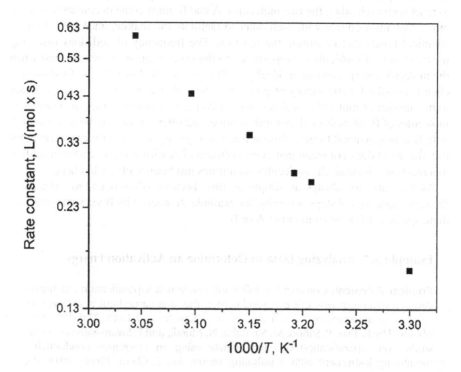

Do these data fall near a straight line, corresponding to the Arrhenius equation? Can you use the data to confirm the reported activation energy of the reaction, 41.4 kJ mol^{-1}? The reaction is overall second-order—how do you know this from the information provided in the problem statement? Given that the reaction is first order in each of the two reactants, can you write a single rate equation that accounts for both the temperature and concentration dependences of the rate?

Example 5.8 Crickets as a Thermometer

Problem statement: In 1897 the following article appeared: A. E. Dolbear, The cricket as a thermometer, *Am. Naturalist*, **1897**, *31*, 970–971. Dolbear found that when a mass of crickets is present together, they chirp in unison, and their rate of chirping depends essentially only on the temperature, provided that the temperature is high enough for them to chirp (about 50°F) and presumably low enough for them to function normally. Here is the equation presented in the publication:

$$T(\text{in }°F) = 50 + \frac{(N - 40)}{4} \tag{5.41}$$

where N is the chirp rate, in chirps per minute.

Analyze the data and determine whether there is a simple equation for the rate of chirping as a function of temperature based on reaction kinetics that might be more fundamentally based than the equation presented in the 1897 publication.

Solution

Let us start by using the reported equation to determine values of N as a function of T. Lacking a form of rate equation and realizing that reaction rate is proportional to the rate constant, we can plot the rate itself (rather than a rate constant) on an Arrhenius-like plot to represent its temperature dependence. Such plots (of reaction rate on a logarithmic scale vs. the inverse of absolute temperature) are often nearly linear. To make such a plot, we express T in K, and plot of the chirping rate, N, on a logarithmic scale vs. $1/T$ (or, more conveniently, $10^3/T$). The data, calculated from Dolbear's published equation, are shown in the table below and in the following plot. We obviously cannot plot the point for 40°F because the rate according to the equation is zero; it is clear that the range of validity of the equation does not extend to a temperature as low as 40°F.

T (°F)	T (K)	$10^3/T$ (K^{-1})	N (chirps/min)
40	277.6	3.602	0
50	283.2	3.531	40
60	288.7	3.464	80
70	294.3	3.398	120
80	299.8	3.336	160
90	305.4	3.274	200
100	310.9	3.216	240

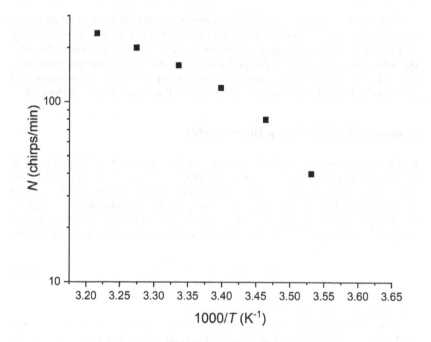

What do we learn from the plot? First, it is not well approximated as a straight line, and we therefore do not represent the data with an Arrhenius-like equation that would represent the chirping rate as exponentially dependent on temperature. Reaction kinetics does not provide much guidance here. Therefore, to use cricket chirping rates as a measure of temperature, we could use the plot with a line through the points as a calibration curve or just use the empirical equation presented by Dolbear.

Example 5.9 Analyzing Kinetics Data for a Second-Order Reaction

Problem statement: Given that the saponification reaction considered in Example 5.7 is first-order in each of the reactants (call them A and B), as determined by the authors of the publication from which the data have been taken, show how to make an informative plot of conversion vs. time data obtained in a perfectly mixed batch reactor for the simple case in which the initial concentrations of A and B are the same.

Solution

With the stoichiometry and equality of concentrations of A and B, the equation for the batch reactor is the following, where k is the rate constant at any given temperature:

$$-r_A = r_B = \frac{-dC_A}{dt} = \frac{dC_B}{dt} = kC_A C_B = kC_A^2 \tag{5.42}$$

Rearranging this, we have

$$dt = \frac{-dC_A}{kC_A^2} \qquad (5.43)$$

Now, to find the time t, we integrate both sides of the equation (some steps in the calculus are left out here; can you check them?), recognizing that k is a constant because the temperature is a constant:

$$\frac{1}{C_A} - \frac{1}{C_{A0}} = kt \qquad (5.44)$$

where C_A is the concentration of A at any given time and C_{A0} is the initial concentration of C_A and of C_B. Thus, if the data agree with the assumed kinetics, a plot of $1/C_A$ vs. t will be well represented by a straight line, with the slope determining k and the intercept determining C_{A0}. Do the data agree with this expectation? Check the source of the data stated in Example 5.7 to find out.

UNRAVELING THE WORKINGS OF A FLOW REACTOR:
ANALYSIS OF A CANDLE AS A CHEMICAL REACTOR

In a classic book, *The Chemical History of a Candle* (1861; republished by Dover, 2002), Michael Faraday reported a number of experiments that showed how a candle works. He showed how to determine essential details of what was in the flame—and where it was in the flame—by sampling the hot gases at various positions in the flame. These experiments included some in which a narrow U-shaped glass tube was held in the flame. The hot gases flowed through the tube, being cooled by the surrounding air, and into a flask below and near the candle. When the tube inlet was placed just above the wick at the center of the flame, vaporized wax flowed into the tube; as it flowed, it was cooled and flowed into the flask where solid wax formed. But when the tube was moved higher up in the flame, a different result was observed: wax was not transported through the tube and into the flask. Faraday figured out something important about how the candle works from the results of these experiments: the wick carries molten wax up from the pool of liquid wax at the top of the candle, and that wax, close to the center of the hot flame, is vaporized, so that, just above the wick, the wax is present in the gas phase and burns there. But farther up from the wick, there is much less vaporized wax—because the wax moving upward in the flame has burned to make CO_2 and water. Faraday observed the flame carefully and noted its different colors and other characteristics at different locations. Can you figure out how to do experiments to show how the concentration of wax varies with axial distance from the wick—that is, how it varies at a given height above the wick?

In other words, the concentrations of the various components (reactants and products) in the flow reactor (the candle) were found to depend on the location in the reactor—both distance from the top of the wick and distance horizontally (radially) from the line extending upward from the wick at a given height above the wick. Thus, the reactor is not a piston-flow reactor. In a piston-flow reactor, the reactant concentration decreases from inlet toward outlet during steady-state operation but does not depend on the radial position at a given distance downstream of the inlet.

Nonetheless, it is clear from Faraday's observations that the concentrations of reactants (wax and oxygen) decline from near the wick to higher above the wick, and, correspondingly, that the concentrations of the combustion products CO_2 and water increase in that direction. These changes contribute to the variations in the flame as a function of position.

Can you think of other simple experiments to help elucidate the workings of the candle as a reactor? Suppose, for example, that the candle is placed in a dish partly filled with water and that a bottle is placed over the candle, with the water providing a seal. After some time, the flame will be extinguished. What can be learned from observations such as the time before the flame goes out and how that depends on the volume of the bottle? Can you figure out how to obtain information about the composition of air from experiments such as these?

CONSTITUTIVE RELATIONSHIPS ARE CONVENIENT, BUT IT IS THE DATA THAT ARE ESSENTIAL FOR REACTOR DESIGN

Solving reactor design problems by combining the appropriate rate equations and design equations, as in examples presented above, is so efficient that it is the typical method. But it is conceptually important to realize that the data providing the basis for determining the rate equations themselves are at least as good as the equations for doing the design—the equations are not needed; only the data from which they can be determined are needed.

To solve the reactor design problems, the typical procedure is to use data from one kind of ideal reactor, combined with the design equation for that kind of reactor plus the design equation for the kind of reactor to be designed. Solving problems this way, without the intermediacy of the rate equation, highlights the basic concepts of the designs.

Suppose, for example, that data have been obtained from a constant-temperature steady-state backmix reactor operated with various feed flow rates and a feed stream of pure reactant A, and further that there is no volume change on reaction. Suppose that a goal is to determine the volume of a steady-state piston-flow reactor with the same feed composition operating at the same temperature, with the feed flow rate and the desired conversion being given.

Measurement of the concentrations of A in the product of the backmix reactor for various flow rates determines values of the reaction rate, as calculated from Eq. (5.21), which is repeated here:

$$r_j = \frac{C_j - C_{j0}}{\dfrac{V}{v}} \qquad (5.21)$$

Note that for each value of C_A in the product, there is enough information to determine r_j from this equation, thus enough information to determine r_j as a function of C_A. To design the piston-flow reactor, we need Eq. (5.26), which is repeated here:

$$r_j = \frac{dC_j}{d\left(\dfrac{V}{v}\right)} \tag{5.26}$$

for the case of $j = A$. To solve, we rearrange the equation to get the following:

$$d\left(\frac{V}{v}\right) = \frac{dC_j}{r_j} \tag{5.45}$$

and recognize that the volume flow rate is a constant, so that we can rearrange the equation to the following for $j = A$:

$$\int_0^V dV = V = v \int_{C_{A0}}^{C_A} \frac{dC_A}{r_A} \tag{5.46}$$

And so to find the reactor volume V, we need to evaluate the integral on the right-hand side of the equation, which requires that we use the data showing how r_A depends on C_A. With a table of data showing this dependence, we could evaluate the integral graphically, whatever the dependence of r_A on C_A. Of course, alternatively, if we had an equation for r_A as a function of C_A (a rate equation), then we could insert that equation into the integral and solve it for V, saving time.

EQUILIBRIUM

Suppose we have a reaction $A \rightarrow B$ that is accompanied by its reverse reaction, $B \rightarrow A$. So far, we have ignored the possibility of the reverse reaction, but, in general, all chemical reactions are reversible, although some may be well approximated as irreversible. Consider the case of a first-order reversible reaction in a constant-temperature batch reactor into which we have introduced A initially. As A reacts to give B, the rate of this forward reaction decreases, and simultaneously, as the concentration of B increases, the rate of the reverse reaction increases. If enough time elapses, the rate of the reverse reaction will match that of the forward reaction. Then the reaction is at *equilibrium*. At equilibrium, forward and reverse reactions occur at the same rate, so that there is no net reaction.

Thus, if we consider Figure 5.4, we realize that the curve bends over eventually to have a slope of zero, corresponding to a net reaction rate of zero. Similarly, if we were to feed A to a piston-flow reactor and measure the concentration of A as a function of y, the distance downstream of the inlet, we would find that the slope of the curve would approach zero as y increased.

If the forward reaction is first-order in A and the reverse reaction is first-order in B, then we can write the net rate as follows:

$$r_{Bnet} = k_f C_A - k_r C_B \tag{5.47}$$

where the subscripts f and r stand for forward and reverse, respectively. Because the net rate of reaction is zero at equilibrium,

$$k_f C_{Aequil} = k_r C_{Bequil} \qquad (5.48)$$

where the subscript equil refers to the equilibrium state.

Rearranging, we find

$$\frac{C_{Bequil}}{C_{Aequil}} = \frac{k_f}{k_r} \qquad (5.49)$$

Now, we know from chemistry that for the reaction A → B, the *equilibrium constant K* is equal to C_{Bequil}/C_{Aequil}, provided that the solution is ideal. Thus, we now have a relationship between the equilibrium constant of the reaction and the forward and reverse rate constants for the reaction, for this special case. Thus, if we measure the forward rate constant and know the equilibrium constant, we can calculate the reverse rate constant as k_f/K (matters are not so simple when the reactions are not first-order).

The idea of equilibrium can be illustrated with both physical and chemical examples. Returning to Chapters 1 and 2, suppose we had two tanks placed on a flat surface, with a horizontal pipe containing a valve connecting the tanks at their bases. Suppose that at time zero, one tank was empty and the other filled with liquid and that the valve was then opened. After some time, the liquid level would become the same in each tank—a state of equilibrium would be reached.

It is essential to recognize that equilibrium and steady state are two entirely different concepts.

CATALYSIS

Most practical reactions and almost all biological reactions are made to go faster by the intervention of catalysts. A catalyst for a reaction increases the rate of that reaction without being consumed in the reaction. Catalysts may be as simple as molecules or ions in solution, but in industrial practice, most catalysts are solids, and the reactions take place on their surfaces. A catalyst that is good for one reaction is usually not good for another reaction unless the two reactions are chemically similar, such as a reaction involving one kind of alcohol and another reaction involving a similar alcohol.

Solid catalysts are advantageous because they are easily separated from fluid-phase products and reactants and because many are stable enough to be used at high temperatures without decomposing. To use catalysts at high temperatures without needing high pressures that would be required to keep reactants and products in the liquid phase (and which require expensive equipment, such as thick-walled reactors), one uses gas-phase reactants, at low pressures.

Catalysts work by making chemical bonds with reactants that give intermediates that react more rapidly to give products than in the absence of catalysts. When the intermediate species formed by combinations of reactants and catalysts react further to give products, the catalyst is regenerated and can work again. Thus, catalysis is a cyclic process; catalysts are used over and over again, to great economic benefit (although they don't last forever and ultimately must be regenerated or replaced). The value of the goods made in the USA in processes that involve catalysis at some stage is several trillion dollars per year—catalysis is the key to efficient chemical transformation.

Because catalysts make reactions go faster, their concentrations appear in rate equations. Consider an example (presented in the work of P. K. Ghosh, T. Guha, and A. N. Saha, Kinetic study of formation of bisphenol A, *J. Appl. Chem.* **1967**, *17*, 239–240, and that of J. I. de Jong and F. H. D. Dethmers, The formation of 2,2-di(4-hydroxyphenyl) propane (bisphenol-a) from phenol and acetone, *Rec. Trav. Chim.* **1965**, *84*, 460–464): the liquid-phase reaction of phenol with acetone gives the valuable industrial product bisphenol A (which is used to make epoxy resins and polycarbonates). This reaction is catalyzed in aqueous solutions of mineral acids such as HCl, and the reaction stoichiometry is the following, where we call acetone A, phenol P, and bisphenol BPA:

$$A + 2P \rightarrow BPA + H_2O \tag{5.50}$$

The form of the rate equation (the kinetics of the reaction) has been found to be the following (and remember that this rate equation does not follow from the stoichiometry and had to be determined experimentally):

$$-2r_A = -r_P = r_{BPA} = \frac{-dC_P}{dt} = kC_A C_P C_{H^+} \tag{5.51}$$

Thus, the reaction is first order in reactant A, first order in reactant P, and first order in H^+, the catalyst, which is an acid such as HCl, with the protons formed by dissociation of the HCl being the catalytically active species. However, under some conditions, the reaction order in P is 2 rather than 1—which illustrates the generally important point that forms of rate equations often change with changes in reaction conditions, such as temperature or solvent or reactant or product concentrations.

This example shows a simple proportionality between the reaction rate and catalyst concentration in a solution, but matters are not always so simple—and remember that it takes experiments to determine how the concentration of the catalyst (or of the reactants or products) affects the rate.

The catalytic reaction of phenol with acetone is one that benefits from a promoter (thioglycolic acid). (A promoter is not a good catalyst for a particular reaction, but it makes a catalyst for that reaction work better.) This point is illustrated by the reaction kinetics (form of the rate equation), where the promoter is represented as T:

$$-2r_A = -r_P = r_{BPA} = \frac{-dC_P}{dt} = kC_A C_P C_{H^+} \left[k_1 + k_2 C_T \right] \tag{5.52}$$

In this equation, k_1 and k_2 are temperature-dependent terms comparable to rate constants. Do you recognize the two limiting cases corresponding to this rate equation? When the second term in the brackets is negligible by comparison with the first (when C_T is small enough so that $k_2 C_T$ is negligible by comparison with k_1), the effect of the promoter on the reaction rate is negligible, and the unpromoted reaction predominates. But when k_1 is negligible by comparison with $k_2 C_T$, the rate of the unpromoted reaction is negligible by comparison with that of the promoted reaction.

Nature's catalysts are enzymes. These are proteins, gigantic molecules (some with molecular weights exceeding a million Daltons) that are synthesized in living organisms by reactions that link amino acids into long chains. Because of their structures and the reactive groups such as COOH groups and OH groups in them that can interact with each other, the enzymes fold into three-dimensional structures that are molecules with slots or clefts where catalytically active groups reside. These can be the OH or COOH groups, for example, or groups that incorporate earth-abundant metals, such as iron or copper. These groups constitute catalytic sites, and they work with great precision because they are the same in each enzyme and placed in precise locations so that they interact specifically with some reactant molecules but not others. Some pharmaceuticals are small molecules that interact specifically with the catalytic sites, and thereby exclude the usual reactant molecules, possibly shutting down key reaction pathways. If these are associated with diseases, the inhibition of one reaction by a pharmaceutical may hinder the progress of the disease. For this reason, chemical engineers work with scientists to understand the diseases, the reactions, and the enzymes and to figure out what pharmaceutical compounds may be effective and how to make them. Such work may involve measurement of rates of reaction in healthy, diseased, and medicated bodies. Thus, catalysis is a central subject that links chemical engineering with sciences including biochemistry.

Many industrial catalysts are solids, used in the form of particles. These are often used in fixed-bed reactors, so that the cost of separation of the products from the catalyst is low—the reactants flow into the reactor, and the products (with unconverted reactants) flow out, with the catalyst particles remaining in place. Solid catalysts are often robust, being used at temperatures as high as about 800 K. To keep the pressures low (and keep the costs down—high pressures require thick-walled equipment, which is often expensive, especially when the material of construction is expensive), the reactants are usually gases. But in many instances, high pressures (high concentrations of reactants) are required for satisfactory reaction rates, and numerous large-scale processes, such as those with hydrogen as a reactant, are carried out at pressures of the order of 100 atm.

Catalysts in solution are less commonly used in technology than solids, because high pressures may be needed to keep dissolved catalysts (and solvents) in the liquid state, because many of them are not stable at high temperatures, and because costs of separating catalysts from products (and recycling them) are usually high. However, nature's catalysts (enzymes) work at low temperatures (e.g., human body temperatures) and are present in the liquid phase (or anchored on surfaces such as those of membranes in biological cells).

REACTION NETWORKS

For simplicity, we have so far emphasized single reactions, but in most industrial reactors, multiple reactions take place simultaneously. Often, many reactions occur simultaneously, even hundreds or thousands of them.

For example, compounds formed by decomposing lignin (from biomass sources such as wood), a potentially important but underutilized raw material for production of chemicals, may be converted into hydrocarbons by catalytic reactions with hydrogen called hydrodeoxygenation reactions. When these reactions take place involving several lignin-derived compounds, a number of products are formed. When these reactions take place together, they are represented in a reaction network. An example of such a network is shown in Figure 5.12. Such a network is a model; in this model,

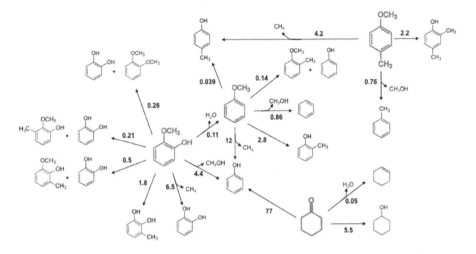

FIGURE 5.12 Reaction network for hydrodeoxygenation of compounds derived from lignin. In the model, each reaction is characterized by a first-order rate constant, in units of L/(g of catalyst×h), with the values shown next to the arrows. Hydrogen is a reactant, but for simplicity is not shown. Nor is each product shown. Reaction temperature was 573 K; the catalyst was platinum nanoparticles supported on high-area porous γ-Al$_2$O$_3$ (gamma-alumina). Reproduced with permission from R. E. Runnebaum, T. Nimmanwudipong, D. E. Block, and B. C. Gates, Catalytic conversion of compounds representative of lignin-derived bio-oils: a reaction network for guaiacol, anisole, 4-methylanisole, and cyclohexanone, *Catal. Sci. Technol.* **2012**, *2*, 113–118.

each reaction is approximated as first order in the lignin-derived reactant. The dependence of reaction rate on the concentration of hydrogen is not shown.

The point about complexity is illustrated by petroleum refining—crude oil contains hundreds or more different compounds (most of them hydrocarbons, which consist of carbon and hydrogen). When these compounds react, hundreds or more products are formed, and many of these react further. Accounting for all these reactions quantitatively is a daunting task, as crude oils from different places have different compositions, and chemical analysis to determine and quantify each compound does not make good engineering sense. Consequently, engineers use approximations to account for the many reactions: they consider groups of compounds (e.g., alkanes with straight chains that have normal boiling points in the range characteristic of gasoline) to be one class of compound. These classes of compounds are called lumps, and approximate reaction networks are determined from experimental results that indicate the rates at which they are converted into other lumps.

An example of such a reaction network of lumps is shown in Figure 5.13. This model reaction network is useful for designing reactors and for predicting the performance of various catalysts for various kinds of crude oils in petroleum refineries. The computations require computers and are routine.

SCALING UP REACTORS AND OTHER EQUIPMENT

In the examples of reactor design that we have considered, it has been a straightforward matter to use the data from any one kind of ideal reactor to design another

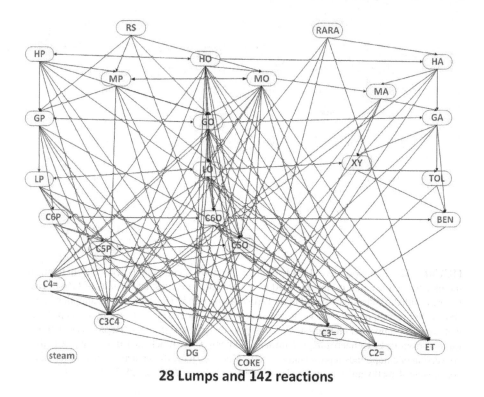

28 Lumps and 142 reactions

FIGURE 5.13 Lumped reaction model approximating conversion of petroleum feeds to a cracking reactor. Each lump represents a group of compounds, which are converted to other lumps in pseudo reactions (referred to as reactions above). The term C4=, for example, represents compounds that are hydrocarbons incorporating four carbon atoms and a C=C bond (these compounds are alkenes, also known as olefins). The simplified representation approximates reactions of hundreds or more compounds to give hundreds or more other compounds. Rate equations for each pseudo reaction are determined by analysis of extensive conversion data. Reproduced with permission from A. Sapre, Role of chemical reaction engineering for sustainable growth: One industrial perspective from India, *AIChE J.* **2022**, *69*, e17685.

ideal reactor. The requirements for the calculations are data characterizing the performance of a reaction in one kind of reactor and the mol balance equations, which are design equations. Thus, we see, for example, how to take data from a small laboratory-scale batch reactor and use them to design a large-scale batch reactor or a large-scale piston-flow reactor. These are examples of *scale-up*—using data from a small piece of equipment as the basis for designing a large piece of equipment.

Often, however, scale-up is not as straightforward as we have implied. The complications arise when matters are more complex than those we have illustrated. Consider this example: A chemical reaction takes place in a batch reactor, and the reaction is exothermic—it gives off heat. We could make the batch reactor almost isothermal—to match the examples considered so far—by removing heat from the reactor as fast as it is generated.

How could we do this? We could insert into the reactor a coiled pipe through which a fluid flowed. The fluid could enter one end of the coil cold and be warmed up by the contents of the fluid in the reactor and exit the coil hotter than it entered. The coil is called a heat exchanger. It could be designed to remove as much heat as is generated in the chemical reaction, thereby making the stirred fluid in the reactor nearly isothermal—how much heat it removes depends on the surface area of the coil that is in contact with the fluid in the reactor and also on the fluid in the heat exchanger and how it flows, among others. We will consider heat exchange in Chapter 6.

Consider the challenge of scale-up of the reactor and the heat exchanger that is essential to the operation of the reactor. The data determined by operation of a small reactor and a heat exchanger in it could be used as a basis for scale-up, but an essential point is that data and the mol balance describing the operation of the reactor are not by themselves sufficient as a basis for the scale-up—data and information about the heat exchanger and how the heat exchanger works are also needed. In other words, the scale-up requires more information than the performance of just the ideal reactor; it requires information about the complete piece of equipment.

One can think of complications beyond what is mentioned in this example that imply that the batch reactor performance data and the ideal reactor mol balance information are not sufficient. Thus, for example, such complications may come into play when the mixing is not perfect. This point raises the question about how the mixing is done. Fluids in reactors can be mixed by stirrers that are simple in design—for example, blades like propellers mounted on a shaft that rotates, driven by an electric motor. Other mixer designs are more complicated—for example, the fluid in the reactor may be pumped internally, and baffles on the reactor walls may facilitate the mixing. The presence of the heat exchanger will affect the flow and therefore the mixing. Predictions of how the mixers depend on the size and shape of the reactor are complex and beyond the scope of what we can consider here.

The essential point of this short section is to point out that scale-up of equipment may be complicated and typically depends on everything that is happening in the equipment. The challenges of scale-up are an important part of engineering design.

RECAP AND REVIEW QUESTIONS

We have now created a basis for design of three ideal reactors. This chapter illustrates the concept of design; with data for a particular reaction occurring in one kind of ideal reactor, we can predict the performance of another kind of ideal reactor (that is, do the design of another kind of reactor). We do this by using the data from the one kind of reactor and the design equation for that kind of reactor, along with the design equation for the other kind of ideal reactor. What connects the two equations (the term that is common to them) is the reaction rate. The reaction rate depends on the temperature and concentrations of reactants (and perhaps of other compounds, such as products and a catalyst and a promoter). It is essential to realize that the reaction rate depends on temperature and concentrations but not on the type of reactor. The molecules do not know what kind of reactor they are in.

Review the three design equations in Table 5.1 for reactions occurring with no volume change. Read these equations, understand them, and visualize them with appropriate graphs. For example, do you understand the close connection between the perfectly mixed batch reactor and the piston-flow reactor? The equations describing them are *essentially the same*—what does this statement mean physically? Do you understand why the conversion in a steady-state backmix reactor (CSTR) is usually less than that in a steady-state piston-flow reactor (when the feed compositions and flow rates are the same)? When is this statement not true? Do you understand why the equations for the perfectly mixed batch reactor and the piston-flow reactor are differential equations, whereas the equation for the steady-state backmix reactor is an algebraic equation?

Can you summarize the essential idea of reactor design when a chemist makes measurements with a perfectly mixed batch reactor and an engineer takes these measurements to find the size of a steady-state piston-flow reactor for the same reaction occurring at the same or another temperature?

This chapter also introduces constitutive relationships that are important in reaction kinetics, and these give equations that facilitate design of reactors. The practically important concepts of catalysts and promoters are also introduced in this chapter, and, because catalysts make reactions go faster, it is no surprise that their concentrations appear in rate equations.

Most biological and industrial reactions are catalytic, and often the catalyst is present as a phase different from the phase of the reactants and products—solid catalysts and gas-phase reactants are especially common in industry.

PROBLEMS

5.1. A cylindrical candle was placed in a completely sealed room initially filled with air at atmospheric pressure, and, as the candle burned, its height was recorded as a function of time with a camera. The following data were obtained showing the candle volume (the candle height times the cross-sectional area) as a function of time.

Time (min)	Volume of Candle (cm³)
0	100
200	55
400	31
600	17
800	9.6
1,000	5.3
1,200	3.0
1,400	2.9
1,600	2.9
1,800	2.9
2,000	2.9

Estimate the rate of burning of the candle as the rate of change of its volume 100 min after the burning started. Estimate the rate of burning of the candle as the rate of change of its volume 160 min after the burning started. Explain in a few words how you would estimate the volume of the sealed room. Make a list of the data you would need to make this estimate.

5.2. The reaction A → B was carried out in an isothermal, perfectly mixed batch reactor, and there was no volume change on reaction. The reactant was present initially with an inert solvent; the only compounds present initially were A and the solvent. The initial concentration of A was 1.0 mol/L. Data from the experiment are shown in the table below:

Time (h)	Concentration of A in Reactor (mol/L)
0.0	1.0
2.5	0.77
6.0	0.60
7.5	0.47
10.0	0.37
12.5	0.29
16.0	0.23
17.5	0.18

A. Consider a batch reactor with three times the volume of the reactor with which the data were obtained, with the initial composition and all the other conditions kept the same. Estimate the rate of reaction r_B in this larger reactor when the concentration of A is 0.26 mol/L.

B. Estimate the rate of reaction r_B in a steady-state CSTR when the feed is A at a concentration of 1.0 mol/L in an inert solvent and the temperature is the same as that in the batch reactor, and the product contains 0.82 mol/L of B.

5.3. The reaction A → B took place at a constant temperature in a steady-state backmix reactor with no volume change on reaction. The reactor volume was 10 L, and the reactor was operated so that the concentration of A in the feed stream was 1.00 mol/L and the concentration of A in the product stream was 0.10 mol/L; the concentration of inert solvent in both streams was 9.40 mol/L. The volume flow rate of the feed was 15 L/min.

A. Determine the volume of a steady-state backmix reactor operated at the same temperature to give the same conversion of A with the same feed when the feed rate is 1,500 L/min.

B. A steady-state backmix reactor with a volume of 1,000 L was used with the same feed and gave a conversion of 90%. What was the volume flow rate of the feed?

5.4. The reaction A → B took place at a constant temperature in a perfectly mixed batch reactor with no volume change on reaction. The data shown in the table below show the concentration of A (and thus the conversion) as a function of time.

A. Use these data to estimate the concentration of A in the reactor at time = zero.

B. Determine values of the rate of change of the concentration of A as a function of time for various times.

C. Make a plot of this rate as a function of time and as a function of the concentration C_A and suggest an equation to represent the data.

Concentration of A (mol/L)	Time (min)
17.0	8.1
14.3	20.1
2.41	95.3
0.81	234
0.48	380
0.30	459

5.5. The data shown in the table below were obtained with an isothermal perfectly mixed batch reactor operated at a constant temperature. The initial composition of the liquid in the reactor was pure compound A, with a concentration of 1.81 mol/L in an inert solvent. There was no volume change on reaction. Use these data to
   A. Determine the rate of the reaction at various values of C_A.
   B. Make a plot of this rate vs. C_A.
   C. Find an empirical equation (constitutive relationship) to represent the data on this plot, assuming that the rate of reaction depends only on C_A.

C_A (mol/L)	Time (min)
1.81	0.0
0.91	40.0
0.45	80.0
0.23	120
0.11	160
0.060	200
0.029	240

5.6. Use the data in the table in Problem 5.5 to determine the volume of a steady-state backmix reactor operated at the same temperature as the batch reactor described in Problem 5.5, given that the feed stream is A with a concentration of 1.81 mol/L in an inert solvent and the product stream contains 0.180 mol/L of A. The volume flow rate is 100 L/h, and there is no volume change on reaction.

5.7. Use the data in the table shown in Problem 5.5 to determine the volume of a steady-state piston-flow reactor operated at the same temperature as the batch reactor described in Problem 5.5, given that the feed stream is A with a concentration of 1.81 mol/L in an inert solvent and the product stream contains 0.180 mol/L of A. The volume flow rate is 100 L/h, and there is no volume change on reaction.

5.8. The data in the following table represent the reaction A → B taking place with no volume change in a perfectly mixed batch reactor at a constant temperature. The initial composition was A with a concentration of 1.00 mol/L in an inert solvent.
   A. Estimate the rate of reaction r_B in the batch reactor when the product contains 0.37 mol/L of A.

Time (s)	Concentration of A in Reactor (mol/L)	Concentration of B in Reactor (mol/L)
0	1.00	0.00
50	0.77	0.23
100	0.60	0.40
150	0.47	0.53
200	0.37	0.63
250	0.29	0.71
300	0.23	0.77
350	0.18	0.82

 B. Estimate the rate of reaction r_B in a steady-state piston-flow reactor operated at the same temperature as the batch reactor when the feed to the reactor is the same as the initial solution in the batch reactor and the product contains 0.37 mol/L of A.

 C. Estimate the rate of reaction r_B in a steady-state backmix reactor operated at the same temperature as the batch reactor when the feed to the reactor is the same as that in part B and the product contains 0.23 mol/L of A.

5.9. The reaction A → B was observed to occur at a constant temperature in a batch reactor initially with a composition of A in an inert solvent with the concentration of A being 2.0 mol/L. There was no volume change on reaction. Data are shown in the following table:

Estimate the volume required for a steady-state backmix reactor fed with A at a concentration of 2.0 mol/L in the inert solvent and operated at the same temperature if the product concentration is 0.51 mol/L and the feed rate is 10 L/h.

Concentration of A (mol/L)	Time (h)
1.70	12.0
1.45	23.2
1.20	48.0
0.80	117
0.51	191
0.31	231

5.10. The following data were determined with an isothermal steady-state backmix reactor with a volume of 30 L for the reaction A → B. There is negligible volume change on reaction. The feed to the reactor was pure A with a concentration $C_{A0} = 10$ mol/L.

Volume Flow Rate of Feed (L/min)	Concentration of B in Product (mol/L)
405	1.0
81	3.0
25	6.0
6.5	7.0
2.5	8.0

Use these data to determine an empirical equation (constitutive relationship) to represent how the rate of the reaction $(-r_A)$ depends on the concentration of A (C_A), assuming that the rate depends on C_A and not on C_B.

5.11. The reaction referred to in Problem 5.5 takes place in a steady-state piston-flow reactor at the same temperature as in Problem 5.5. The feed to the reactor contains A in an inert solvent, at a concentration of 1.0 mol/L. Use the data in the table determined with a backmix reactor to find the residence time required in the piston-flow reactor for the concentration of A to decline to from 1.0 to 0.10 mol/L.

5.12. The reaction of ethyl acetate $(CH_3COOC_2H_5)$ with a base takes place in aqueous solutions:

$$CH_3COOC_2H_5 + NaOH \rightarrow CH_3COONa + C_2H_5OH$$

This is called a saponification reaction, and data characterizing its kinetics are presented in one of the problems in this chapter. Find out what the word "saponification" means and look up the applications. What products are manufactured in saponification reactions? Using information in this chapter, design an isothermal batch reactor to make 1,000 kg of product sodium acetate (CH_3COONa) assuming that the temperature is 330 K and that the initial concentrations of ethyl acetate and sodium hydroxide in the reactor are each 3.00 mol/L and that the ethanol (C_2H_5OH) formed in the reaction is removed continuously by vaporization, with other components in the reactor condensed in a condenser and returned continuously to the reactor. Neglect the volume change on reaction as a first approximation. What is the approximate reactor volume (and what information do you need to determine this)? How long does it take to convert 99% of the ethyl acetate? Explain why the rate of the reverse reaction can be neglected in the analysis.

5.13. Construct a problem asking students to design a piston-flow reactor for the saponification reaction mentioned in Problem 5.12 using data presented in this chapter.

5.14. The frequency of flashing of fireflies, like the frequency of chirping of crickets, is suggested to provide a measure of temperature. Find data that allow you to evaluate this suggestion.

5.15. Reactions often occur simultaneously, and the term reaction network (or reaction scheme) is used to represent them. For example, the reaction A → B may occur at the same time as the reaction A → C and/or with the reaction B → C.

A. Represent these reactions schematically.
B. Figure out which reactions are parallel and which are sequential.
C. Go to the literature and find publications by M. T. Klein and K. B. Bischoff and prepare a summary of the method they call the Delplot method to resolve such reactions.

5.16. The reaction network shown in Figure 5.12 is simplified; the full stoichiometry of each reaction is not shown. Figure out what the stoichiometry of each reaction is and write it.

5.17. The reaction network shown for lumps in Figure 5.13 is simplified. Pick one pseudo reaction as an example (e.g., a reaction of C4=), and figure out the full stoichiometry of one of the simpler pseudo reactions (a set of reactions) and write them, including the full representation of the compounds that are reactants and those that are products. Realize that a lump such as C4= represents more than one compound (in this case, the isomers of butene, of which there are four). Solving this problem requires looking up some organic chemistry.

5.18. Find out what bisphenol A is, what it is used for, and why it is hazardous to human health.

5.19. Most solid catalysts used in large-scale processes in industry are porous particles—some have pores of the order 1 nm in diameter and internal surface areas of the order of $100\,m^2/g$. Catalyst particle diameters are of the order of 1 cm. Reactant molecules diffuse in the pores to reach the interior surface. When the reactions are extremely fast, and/or when the diffusion in the pores is extremely slow, and/or when the catalyst particle diameter is extremely large, the reactant molecules may be mostly consumed by reaction before many of them get close to the center of a particle. Then, the catalyst is not used efficiently—parts of the internal surface are not accessible enough for the reaction to be as fast there as near the pore mouths. Suppose that the reaction, when it occurs in the absence of diffusion influence, is first order, characterized by a rate constant k. Further suppose that the catalyst particles are spherical, with radius R and that the diffusion of reactant molecules is characterized by a diffusion coefficient D. Find a dimensionless group that you expect to be important in the analysis of the reaction with diffusion process.

6 Energy, Energy Balances, Heat Transfer, and Temperature Control

ROADMAP

A principle that is as important as the conservation of mass is the principle of conservation of energy. This chapter is about energy in its various forms and about analysis and design when energy transfer is important, such as when a hot rock is placed in cold soup, heating the soup (which gains energy) as the rock is cooled (losing energy). Phenomena such as this are analyzed beginning with energy balances. We apply them with constitutive relationships, introducing the heat capacity and illustrating heat transfer and associated changes in energy with changes in temperature. We consider heat conduction and introduce thermal conductivity to analyze the phenomena and introduce calorimeters and their applications to determine heats of chemical reactions and heats of fusion, vaporization, and solution. This final chapter integrates ideas and methods from the earlier chapters and is intended to illustrate the connections between them and display more of the diversity of chemical engineering.

What Is Energy?

A central engineering principle is that mass is conserved—not created and not destroyed. Equally important is the principle that energy is conserved. Engineers often begin analysis of a physical situation with an energy balance, just as they often begin with a mass balance. Sometimes the two are used together, or in conjunction with a mol balance, when chemical reactions take place.

What is energy? Think of atoms and molecules in a fluid and the influence of temperature on them. The higher the temperature, the more energy a molecule has and the faster it moves in a fluid, separate from any bulk flow of the fluid. Moving molecules *transfer* energy to other molecules by colliding with them: the faster molecule transfers energy to the slower (lower-energy) molecule in the collision. The energy of a molecule is associated not just with its movement within a fluid, but also with movements of the atoms within the molecule—including vibrations, rotations, and twists. The higher the temperature, the more energetic and rapid are these movements. Think of a simple molecule that consists of only two atoms, carbon monoxide, CO. The C and O atoms are bonded to each other, and they constantly vibrate, moving closer and then farther from each other like a vibrating spring, with the C–O bond remaining intact unless the conditions are severe, such as an extremely high temperature. The higher the temperature, the higher the vibrational frequency.

DOI: 10.1201/9781003429944-6

Electrons in the atoms also have energy associated with their motions. In solid metals, electrons move rapidly between atoms, carrying negative charge, so that the metals are good conductors of electricity. The rapid electron movement also makes the metals good conductors of energy. We often refer to this energy as heat. In other words, metals are good heat conductors.

Because chemical engineers work with bulk samples and systems that contain many atoms and molecules, they need to account for forms of energy in addition to those internal to the atoms and molecules; this additional energy is associated with the bulk movement of fluids (or solids)—and also the potential for this bulk movement. These latter forms of energy *need to be accounted for separately from the internal energy associated with the local movement of molecules, atoms, and electrons within a fluid or solid.*

In considering the energy of a bulk fluid, we make a distinction between mechanical energy associated with the flow, such as the energy of water flowing through a turbine that causes it to turn. This mechanical energy (or kinetic energy, a term that has essentially no connection to reaction kinetics as we used that term in Chapter 5) can result from the conversion of potential energy. This point is illustrated by the conversion of potential energy of water stored in a reservoir into kinetic energy of water flowing downward to a turbine and forcing it to turn. The higher the level of the water above the turbine, the greater the potential energy it has—that is, the greater its potential to drive the turbine.

In the examples we worked on with draining tanks, we saw potential energy converted into kinetic energy as water flowed from a tank down through an orifice. The higher the level of liquid in the tank, the greater its potential energy—and the faster the flow of the liquid out of the tank (the greater the kinetic energy). To repeat, in fluid flow driven by gravity, potential energy of the fluid is converted into kinetic energy.

Changes in the energy of a bulk fluid may lead to changes in the internal energies associated with the atomic and molecular motions. Thus, for example, when water flows downward from a reservoir and through a turbine, its temperature may change, so that its molecules move and vibrate at rates different from before. If we hold a glass bottle of carbonated water—with the liquid containing dissolved carbon dioxide at a pressure greater than atmospheric—and then crack open the cap, we see bubbles form, hear carbon dioxide gas flowing out as the pressure in the bottle decreases, and feel the bottle getting colder as a result of the process. (This temperature change is associated with a phase change—liquid to gas—and we consider this topic below.)

We refer to the total energy as the sum of all these kinds of energy: the internal, kinetic, and potential energies. Other kinds of energies are important too (such as magnetic and electrical energies), but to keep things simple and focus on topics most common in chemical engineering, we mostly neglect these other forms of energy in this book.

The presentation here is far simpler than what students will encounter later in chemical engineering subjects such as thermodynamics.

ENERGY CONVERSION TECHNOLOGIES

Energy conversion is central in our technological world and practiced on an enormous scale. The technologies include hydroelectric power generation, whereby the potential energy of water is converted to kinetic energy and then electrical energy in turbines. As wind turns a turbine, it generates electrical energy. Huge amounts of energy are converted as coal, petroleum, and natural gas (a mixture of methane with small amounts of ethane and propane) are burned. The highly exothermic combustion reactions of these fossil fuels generate hot gases (mostly carbon dioxide and water, as in candle burning) that are used to heat liquid water and convert it into steam in power plants. The steam is used to drive turbines and make electrical energy.

A schematic representation of the overall process of coal conversion into electricity is shown in Figure 6.1. Carbon dioxide, a greenhouse gas generated in fossil fuel burning, is produced on an enormous scale in such facilities, enough to have substantially raised the temperature of the earth's biosphere and to have upset its climate and ecology—a vital concern to all of us inhabiting our planet and a responsibility that chemical engineers face—and are uniquely qualified to address.

Environmental concerns are driving rapid changes in energy conversion technologies, and conversion of sunlight into electricity with photovoltaic cells is a large and growing technology, as is the conversion of wind energy into electricity. Targets for the future include new technologies for using sunlight to provide the energy for chemical reactions.

MEASURING TEMPERATURES

The higher the temperature of a gas, liquid, or solid, the higher its internal energy. Temperature is thus a physical property that we need to account for in energy conversion. We can measure temperature. Temperature scales are defined on the basis of standard temperatures, those at which well-known phase transformations take place: the freezing of water at atmospheric pressure and the boiling of water at atmospheric

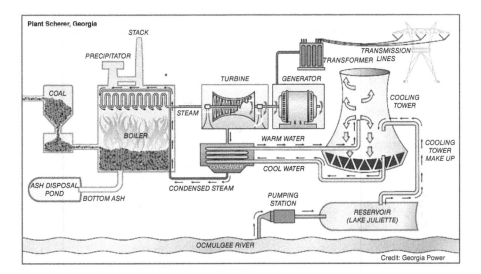

FIGURE 6.1 Schematic representation of a coal-fired power plant showing how the heat generated by coal burning is converted into electrical energy. Notice how the technology influences the environment: carbon dioxide and other gases and solid waste are emitted; water is taken from a river. *Source*: Reproduced with permission from usgs.gov, an official website of the U.S. government.

pressure. The freezing point of water is defined as exactly 0°C and its boiling point as exactly 100°C—these are the anchor points of the Celsius temperature scale. The anchor point for the Kelvin scale is 0 K, absolute zero—the lowest temperature that could be reached, and the Kelvin scale, as we have seen, is linked simply to the Celsius scale, by a difference of 273.2 degrees at any temperature.

We learned in Chapter 5 that the chirping rate of a group of crickets is a measure of temperature; the crickets are a thermometer. Common thermometers take advantage of the temperature dependence of liquid densities. Mercury has been used often as a thermometer fluid. Its density (Table 6.1) depends almost linearly on temperature over a wide range (Figure 6.2). This property makes mercury a good thermometer fluid.

TABLE 6.1
Densities of Liquid Mercury at Various Temperatures

Temperature (K)	Temperature (°C)	Density (kg/m³)
273.2	0	13,595
293.2	20	13,545
323.2	50	13,472
373.2	100	13,351
423.2	150	13,231
473.2	200	13,112
523.2	250	12,993

Source: www.EngineeringToolbox.com, accessed May 23, 2022.

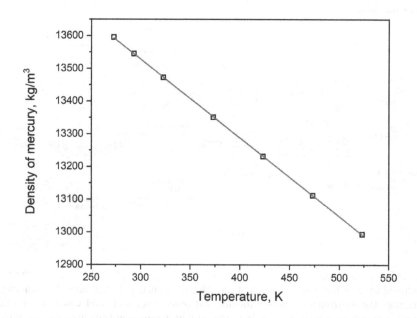

FIGURE 6.2 Temperature dependence of density of liquid mercury at atmospheric pressure (data from Table 6.1).

Example 6.1 Design of a Mercury Thermometer

Problem statement: Using the data presented in Table 6.1, design a wall-mounted laboratory thermometer with mercury as the thermometer fluid.

Solution

Mercury offers good physical properties for a thermometer, being liquid over a wide range of temperatures and having a nearly linear dependence of density on temperature (Figure 6.2). Let us assume that the lowest temperature reading will be -30°C and the highest 150°C. According to Table 6.1, the density of mercury is 13,234 kg/m³ at 150°C and, by extrapolation, 13,666 kg/m³ at -30°C. These data show that the density increases by 3.3% as the temperature increases from -30 to 150°C. This result is a starting point for the thermometer design.

Here are some suggested design criteria: The thermometer should have a scale with numbers that can be read conveniently that indicate the volume of the mercury. Placement of all the mercury in a container such as a cube does not give a practical design, because the volume of mercury needed to give a conveniently measured change in the liquid height in the cube is large (can you calculate how large?) and mercury is toxic—so that we should minimize its volume. We need instead a design that makes the change in mercury volume easily readable over the design temperature range. Thus, we suggest a spherical bulb connected to a narrow column mounted vertically on top of it, with the volume of mercury only slightly exceeding the volume of the bulb at the lowest temperature. With this design, mercury will rise in the column as the temperature increases. We suggest a column height of 15 cm to give an easily read temperature scale. We suggest a scale in °C on one side of the column and °F on the other, because many people are familiar with one or the other of these scales. If we assume a negligible volume in the cylinder above the bulb and below the reading at -30°C, we can calculate the volume of the column for a given bulb volume.

Let us assume a glass bulb with a volume of 2.0 cm³; the additional volume of a glass column above it is chosen on the basis of the statement above to be 3.3% of that volume, or 0.065 cm³. With the assumed cylinder length of 15 cm, the cross-sectional area of the inside of the cylinder (volume divided by length) is calculated to be about 0.037 cm, small and not as easily visible as we would prefer. A wider, flatter space inside the cylinder seems to be a better option.

We have implicitly made a simplifying assumption in doing this design: we neglected the temperature dependence of the volume of the glass and the material to which it is mounted. Can you estimate how much error would be associated with the neglect of these temperature dependences?

Mercury thermometers have a long history of providing accurate temperature measurements, but, because of the mercury's toxicity, many have been replaced by safer instruments.

Other devices that engineers commonly use to measure temperatures are thermocouples. A thermocouple is a sensor that consists of a junction of two wires of different metals that are welded to each other; the junction is typically of the order of a millimeter in diameter. Because the metals are different, a voltage is generated near the junction; this is a consequence of a physical phenomenon known as the Seebeck effect—it occurs just because the metals are different from each other. The voltage depends both on the temperature and on the types of wire used in the device. The wires are usually metal alloys (e.g., chromel and alumel).

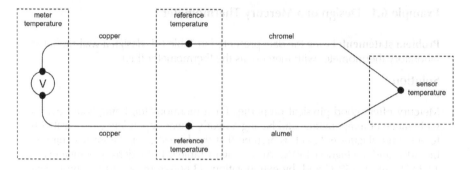

FIGURE 6.3 Electrical circuit showing how a thermocouple works for measurement of temperature. The symbol V indicates an instrument for measuring and recording voltage.

Thermocouples are robust and give continuous temperature readings. They are widely used for measuring temperatures over wide ranges.

Thermocouples are used in electrical circuits like that shown in Figure 6.3. Typical voltages are of the order of microvolts. Determination of the temperature of the junction requires measurement of three voltages in the circuit: those of the thermocouple junction and those of two reference junctions, which provide calibrations by being placed in surroundings of known temperature, such as ice–water mixtures at equilibrium. Tables of data converting thermocouple voltage readings to temperatures are widely available, for example, from the U.S. National Institute for Standards and Testing (NIST).

HEAT TRANSFER: HEATING AND COOLING SAMPLES BY CONDUCTION, RADIATION, AND CONVECTION

Matter in any form can be heated by addition of energy and cooled by removal of energy. The energy (heat) can be transferred by the mechanisms mentioned in Chapter 1: conduction, radiation, and convection. These mechanisms of heat transfer often occur at the same time—as in the boiler in Figure 6.1.

Temperatures can be controlled by regulation of the rates of heat transfer (heat input or removal from a system). Control of heat transfer rates for control of temperatures is illustrated in nature by cold-blooded animals. They adjust the rates of heat transfer to or from them by moving between sun and shade and from above ground to below ground. Further, they use subtle methods for fine control of their temperatures by simultaneously taking advantage of radiation, conduction, and convection.

Lizards basking in the sun are warmed by radiation. To maximize their heat absorption to warm up quickly and have full mobility to seek prey and mates and avoid predators, some lizards take on darker colors as they are heating up to become more efficient heat absorbers. These animals must avoid becoming too hot, because overheating is fatal, and so when they have absorbed enough radiation, they may become lighter in color and reflect more energy and absorb less energy (Figure 6.4).

FIGURE 6.4 Namaqua chameleon (*Chamaeleo namaquensis*) in the Namib desert in Namibia demonstrating various colors and color patterns for thermoregulation. The images in this figure represent a single animal in its light phase (top), dark phase (middle), and combined intermediate phases (bottom). The top and bottom images show a combination of the light phase on one side and a slightly darker phase on the other. Looking at the bottom image, can you see which side of the chameleon is absorbing radiation and figure out whether it was maximizing or minimizing absorption of radiation by having the left side darker than the right side?

FIGURE 6.5   Wedge-snouted skink (*Trachylepis acutilabris*) in Namibia minimizing conductive heat transfer from hot ground by raising its tail and feet to minimize the area of contact with the ground.

To be warmed by conduction, a lizard may flatten itself on a hot rock to increase the area for heat transfer from the rock. When it is too hot, the lizard may minimize absorption of heat by conduction from a hot surface by lifting its tail and feet to minimize the area of contact through which heat is conducted (Figure 6.5). To cool itself, the animal may stand at a high point and straighten its legs to expose a maximum of its surface to flowing air for cooling by convection (Figure 6.6).

Animals use all these mechanisms of heat transfer simultaneously and often control their temperatures within ranges of only a few degrees or less when they are active. When they are too cold, they move sluggishly and are susceptible to predation. When they brumate (cold-blood animals brumate; warm-blooded animals hibernate), they are sluggish and inactive and need to be where predators will not find them. During brumation, they become much colder than when they are active, and their metabolisms slow down so that they use only little energy to stay alive.

Heat Capacity and Energies of Heating and Cooling of Solids, Liquids, and Gases

To understand how much the temperature of a fluid or solid changes when heat is added or subtracted, we need a physical property of the fluid or solid, its heat capacity—and a constitutive relationship. The heat capacity, represented by the symbol C_p (the subscript p refers to the value at a constant pressure) of a given gas, liquid, or solid, is defined as the amount of heat required to change its temperature by one degree; common units of C_p are J/(kg K) and kJ/(kg K).

When heat is added or subtracted from a sample, the amount of energy transfer is thus $mC_p\Delta T$, where m is the sample mass and ΔT is the temperature change. (For simplicity, we have assumed that the temperature is the same throughout the sample and

FIGURE 6.6   Greater earless lizard (*Cophosaurus texanus*) in Arizona maximizing convective heat transfer to the air by maximizing exposed body area to air currents while also minimizing conductive heat transfer from the hot rock by raising parts of its hind feet.

that the heat capacity is independent of temperature.) For example, knowing that the value of C_p for liquid water at atmospheric pressure is about 4.2 J/(g K) (Table 6.2), we can calculate how much heat is required to raise its temperature from one value to another. Thus, the energy required to heat 100 g of liquid water from 273.2 K to 323.2 K is $(100 \text{ g}) \times (50 \text{ K}) \times [4.2 \text{ J/(g K)}] = 2.1 \times 10^4 \text{ J}$.

Values of C_p are available for many gases, liquids, and solids, as exemplified by the data in Table 6.2. Heat capacities depend on temperature, and we have simplified matters by providing no information about the temperature dependences in this table.

Phase Transitions and Energy Changes Associated with Them

When a compound such as water at a constant pressure undergoes a phase change, such as from solid to liquid or from liquid to gas, substantial energy changes are needed, even with the temperature remaining unchanged. The energy changes are called the heat of vaporization as a gas forms from a liquid (heat of condensation accounts for the reverse; the value is the same, but the sign is opposite) and the heat of fusion as a liquid forms from a solid (or vice versa). These terms are separate from heat capacity terms and not related to heat capacity.

TABLE 6.2
Heat Capacities at a Constant Pressure of Some Gases, Liquids, and Solids

Compound or Mixture	State (Gas, Liquid, or Solid)	C_p, kJ/(kg K)
Air	Gas	1.01
Carbon monoxide	Gas	1.02
Carbon dioxide	Gas	0.844
Methane	Gas	2.22
Water (steam)	Gas	1.97
Ethyl alcohol	Liquid	2.3
Methyl alcohol	Liquid	2.5
n-Propyl alcohol	Liquid	2.4
Water (at 293 K)	Liquid	4.19
Copper	Solid	0.39
Silver	Solid	0.23
Glass (plate)	Solid	0.5
Water (ice) (at 273 K)	Solid	2.09
Quartz (SiO_2)	Solid	0.73

Source: www.EngineeringToolbox.com, accessed June 25, 2022.

TABLE 6.3
Heats of Fusion and Heats of Vaporization of Some Compounds. The Data Represent Changes at Atmospheric Pressure

Compound	Heat of Fusion at Normal Melting Point (kJ/kg)	Heat of Vaporization at Normal Boiling Point (kJ/kg)
Ethyl alcohol	108	846
Methyl alcohol	98.8	1100
n-Propyl alcohol	86.5	779
Copper	206	
Mercury	11.4	295
Nickel	293	
Silver	105	
Water	334	2256

Source: www.EngineeringToolbox.com, accessed June 9, 2022.

If we convert liquid water at 50 K above its freezing point to solid water at 10 K below its freezing point, for example, then to calculate the energy change we need to know the heat capacity of the liquid to account for the cooling from 50 K above the freezing point to the freezing point; plus the heat of fusion as the liquid becomes solid at the freezing point; plus the heat capacity of the solid to account for the cooling from this temperature to 10 K below the freezing point.

Heats of fusion and heats of vaporization are much greater for some compounds than for others, as shown in Table 6.3.

Example 6.2 Energy Required to Heat Ice at a Temperature Below Its Melting Point to Steam at a Temperature Above the Normal Boiling Point of Water

Problem statement: Using the data presented above, calculate the energy required to heat one kilogram of water at atmospheric pressure from 50 K below its melting point to 50 K above its melting point and then further to 50 K above its boiling point. The changes involve heating of ice, melting of ice, heating of liquid water, vaporization of liquid water, and heating of steam.

Solution

The normal freezing point of water is 0°C (273.2 K), and the normal boiling point is 100°C (373.2 K). Several terms are needed to calculate the energy required to carry out the process described: (1) heating the ice from 223.2 K to the melting point, (2) heating the ice at its melting point to convert it entirely to liquid water at this temperature, (3) heating the liquid water from this temperature to its normal boiling point, (4) heating the liquid at this temperature to create steam at this temperature, and (5) heating the steam from this temperature to the final temperature of 423.2 K. To calculate these values, we need the mass of water (one kilogram), the heat capacity of ice, the heat of fusion of ice, the heat capacity of liquid water, the heat of vaporization of water, and the heat capacity of steam.

Taking values for water of the heat capacities from Table 6.2, the value of heat of fusion from Table 6.3, and the value of the heat of vaporization from Table 6.3, we add the terms $mC_p(\Delta T)$ for ice [(1 kg)(2.09 kJ/kg K)(50 K)], the heat of fusion of the water [(1 kg)(334 kJ/kg)], the $mC_p(\Delta T)$ term for liquid water [(1 kg)(4.19 kJ/(kg K)) (100 K)], the heat of vaporization of the water [(1 kg)(2256 kJ/kg)], and the $mC_p(\Delta T)$ term for steam [(1 kg)(1.97 kJ/(kg K))(50 K)], finding a total of 3212 kJ. Note how the term for the heat of vaporization of water is much larger than the others. Can you figure out why?

Temperature Control by Melting and Vaporization

Heats of fusion and heats of vaporization are used to advantage in many designs. For example, desert water bags illustrate the benefits of evaporative cooling and the relatively large value of the heat of vaporization of water. These bags are made of linen, flax, or other materials consisting of fibers that hold almost all the water inside the bag while letting small amounts diffuse between the fibers. When the water reaches the outside surface of the bag, it evaporates in the hot desert environment, removing energy from the bag and its contents and cooling them.

These bags have become antique collectibles and are regarded by some people as works of art. Imagine the bag outside the window of a moving car, so that the convective motion of the air increases the rate of evaporation of the water and thereby the rate of cooling of the water. People used these bags to cool water for car radiators in years past when radiator overheating was common in hot deserts.

Evaporative cooling is also used by animals, such as the desert gecko shown in Figure 6.7. Water is precious and normally retained with high efficiency by such

FIGURE 6.7 *Pachydactylus scherzi* with mouth agape to cool itself by evaporation of water from its open mouth. This desert-dwelling gecko is nocturnal and had just been exposed to the hot sun by removal of a rock that it was hiding under. With a closed mouth, the animal loses very little water by evaporation. Using information presented in Chapter 1, can you estimate the area through which water would be transferred by evaporation from the gecko?

animals, but, in an emergency, to avoid dangerous (even fatal) overheating, the animal opens its mouth to be cooled by evaporation of water. We humans also benefit from cooling by evaporation of water (perspiration). Can you explain why such cooling is more efficient in an environment with low humidity than in one with high humidity? And why we feel dry in the desert on a hot day and wet in a rain forest on a hot day?

The heat of fusion of water accounts for the effectiveness of household ice boxes. Ice in a container can keep food at near the melting point of water for a day or more, provided that the container is well insulated and not opened too often. Thus, the temperature of the ice box is *controlled* as long as the ice has not all melted—the temperature in the box becomes nearly the temperature at which ice melts.

Ice boxes have found wide application in houses without electricity, and, years ago, it was common in the U.S. and other countries for people to receive regular, even daily, deliveries of blocks of ice with masses of about 50 kg each, insulated during transportation, for example, by packing in straw or sawdust. Ice was commonly transported by horse-driven carriages. In the winter, in some regions such as New York and New England, the ice was harvested from ponds and rivers. Blocks of ice were transported in insulated vehicles, sometimes for long distances, and stored in ice houses, some placed underground for good insulation. Above-ground ice houses were painted bright white (can you figure out why?)

A photograph of an antique household ice box is shown in Figure 6.8. Now we consider its design.

Example 6.3 Design of an Antique Icebox

Problem statement: Design an ice box for household use to contain a single block of ice with a mass of 40 kg.

FIGURE 6.8   Antique household icebox having a height of about 1.2 m.

Solution

The image in Figure 6.8 provides a starting point, with an icebox of an appropriate size. Insulation will have been included in the wall; in old ice boxes, the insulation might have been straw, sawdust, or cork. Today, we would choose a material such as fiberglass—a discussion of insulators is presented later in this chapter. The icebox would need a sturdy tray large enough for the block of ice (placed near the bottom, so that water formed by melting would not flow over the food). A tray would be used to collect the water, and a drain would be helpful. Separate sliding trays would hold the food and beverage containers. The doors would be sealed well, likely with rubber gaskets. Overall, the design is not much different from that of a modern electric refrigerator.

WINDOW DESIGN FOR ENERGY CONSERVATION IN BUILDINGS

Chemical engineers are applying some of the above-mentioned ideas about heat transfer to design improved windows for office buildings and homes. The window materials are chosen to transmit almost all visible light but to limit the transmission of high-energy near-infrared light in the summer, but not in the winter. The materials in these windows are expensive, and applications have not yet become common, but the incentive for saving energy by minimizing that needed for heating and cooling buildings provides a strong motivation to improve the designs and make them less costly.

For temperature control, windows could be made of materials that, like some lizards, adapt to the conditions—when it is too hot, the windows would become tinted to restrict light transmission and keep rooms cooler, and when it is too cold, the tint would disappear and maximize the entry of natural light to help heat the interior space. Such windows are already used in boats, cars, and airplanes and are finding initial applications in public buildings and office buildings. Because buildings are responsible for a large fraction of the energy consumption in the U.S. (in heating, cooling, and ventilation) and because a large fraction of the heat transferred into and out of buildings is through the windows, windows that control heat transfer seem destined to have a large impact on energy conservation.

TEMPERATURE CONTROL WITH AZEOTROPES

We have seen in Chapter 3 how mixtures can be separated by distillation, by taking advantage of the differences in boiling points of the solution components. But some fluid mixtures have properties (related to the strengths of interactions between the molecules) that limit this approach. For example, it might at first seem as if one could separate ethyl alcohol and water by distillation because these two compounds have substantially different boiling points at atmospheric pressure (normal boiling points): 351.5 and 373.2 K, respectively. But a solution of about 95.6% ethyl alcohol and 4.4% water (by mass) boils at a temperature 0.2 K less than that of the pure alcohol, and so a solution of this composition boils as if it were a pure compound. That is, when the solution is richer in water than this composition and is brought to a boil, the vapor that is formed has the composition 95.6% ethyl alcohol and 4.4% water by mass—it is not pure ethyl alcohol. Thus, separation of such water-rich mixtures by distillation is not possible. Consequently, ethyl alcohol–water mixtures of this composition are commonly made and marketed, and, if purer ethyl alcohol solutions are needed, other methods of separation are needed.

This composition is called the azeotropic composition, and the mixture is called an azeotrope or an azeotropic mixture. (In distillation, adding a third component, such as benzene, can change matters—ethyl alcohol of higher purity than the azeotrope can be made, but it will contain some benzene because the separation will not be perfect.)

Many combinations (pairs) of compounds form azeotropes. We can take advantage of them as temperature controllers—we can find ranges of compositions of some mixtures that all boil at the same temperature.

The following is an example of how an azeotrope facilitates a set of experiments to determine reaction rates. As was shown in Chapter 5, rates of reactions depend strongly on temperature, and experimentalists often make measurements of reaction rates at separate, controlled temperatures to resolve the dependence of rate on temperature.

Example 6.4 Design of a Reactor for Measuring Rates of a Chemical Reaction in Azeotropic Solutions Used for Temperature Control

Problem statement: Design a simple reactor for measurement of the kinetics of the reaction of *tertiary*-butyl alcohol to give isobutylene and water catalyzed by particles of sulfonic acid ion-exchange resin at a controlled temperature.

Solution

Looking up information about *tertiary*-butyl alcohol (TBA) and water, we find that they form an azeotrope consisting of 88.2 mass% TBA and 11.8 mass% water that boils at 353.1 K (X. Songlin and W. Huiyuan, Separation of tert-Butyl Alcohol-Water Mixtures by a Heterogeneous Azeotropic Batch Distillation Process, *Chem. Eng. Technol.*, **2006**, *29*, 113–118). TBA has a normal boiling point of approximately 356 K, and so we can use a range of compositions with lower TBA concentrations and expect them to have the same temperature when they are boiling. Recognizing that isobutylene is much more volatile than either TBA or water (its normal boiling point is about 266 K), we see how it can be separated as a gas from the reaction products and how its rate of formation can be measured with a soap-film flow meter, as follows.

The reaction stoichiometry is the following:

$$C(CH_3)_3 OH \rightarrow C(CH_3)_2 (CH_2) + H_2O \qquad (6.1)$$

The suggested reactor design is shown in Figure 6.9: it allows measurement of the rate of reaction as the rate of flow of isobutylene gas, as the water formed in the

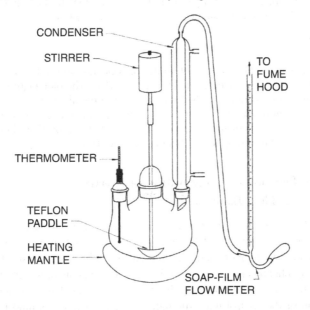

FIGURE 6.9 Reactor designed for measurement of kinetics of reaction of *tertiary*-butyl alcohol to make isobutylene and water, with the temperature controlled by the *tertiary*-butyl alcohol–water azeotrope, which boils as the rates of isobutylene formation are measured with the soap-film flow meter. The stirrer provides good mixing of the liquid and the catalyst particles that are suspended in it. As the catalytic reaction proceeds, isobutylene gas is formed and escapes from the liquid, flowing to the condenser. The boiling also creates water and alcohol vapors. The alcohol and water are converted to liquid in the condenser and return to the reactor, but the isobutylene has such a low boiling point that it flows through the condenser—to the soap-film flow meter, where its flow rate is determined. This reactor is called a semi-batch reactor, because some but not all of the components remain in the reactor and there is no flow into the reactor (except for the recycled fluids from the condenser). Figure reproduced with permission from B. C. Gates and J. D. Sherman, Experiments in Heterogeneous Catalysis: Kinetics of Alcohol Dehydration Reactions, *Chem. Eng. Educ.*, **124**, summer 1975.

reaction remains in the stirred reactor. Thus, as the reaction proceeds, the vaporized TBA and water are condensed in the condenser and returned to the stirred reactor with the boiling liquid and the particles of catalyst. The isobutylene gas that forms is too volatile to be condensed in the condenser, and the rate of the reaction is determined with the soap-film flow meter as the flow rate of isobutylene gas. As the reaction proceeds, the liquid becomes depleted in TBA and richer in water, and so we extrapolate the rate data to time = zero to determine the reaction rate that corresponds to the initial composition.

Example 6.5 Determining Kinetics of a Reaction Using Data from the Reactor Design Shown in Example 6.4

Problem statement: Data characterizing the reaction stated in the preceding example are shown in Table 6.4. Rates were found to be proportional to the mass of catalyst suspended in the liquid in the reactor, and so rates are reported per catalytic site (the sites are sulfonic acid groups in the solid catalyst). Rates of reaction were measured as volume flow rates of the gas and are reported with the dimensions shown in the table. The data are from B. C. Gates and W. Rodriguez, General and Specific Acid Catalysis in Sulfonic Acid Resin, *J. Catal.*, **1973**, *31*, 27.

Solution

To begin analyzing the data, we plot rate vs. concentration of the alcohol to find out whether a simple equation will represent the data. The plot (Figure 6.10) shows that the rate is proportional to the concentration of the alcohol, and so we have a first-order reaction. The slope of the line determines the rate constant in the equation $r = kC_{TBA}$. We calculate a value of k of 1.8×10^{-4} L/(mol of SO_3H groups × s) from the slope of the straight line that best fits the data.

APPLYING THE PRINCIPLE OF CONSERVATION OF ENERGY TO ACCOUNT FOR HEAT TRANSFER

As we did in applying the principle of conservation of mass, we apply the principle of conservation of energy by first defining a system, such as that pictured in Figure 3.1. Statements of conservation of energy are the starting points for analysis of many chemical engineering phenomena. The system is often the contents of a container. A control mass is contained in a volume defining the system. The dashed lines in Figure 3.5 denote the movement of the control mass through the fixed volume. As fluid moves through this control volume, the total energy of the system can change in a variety of ways. For example, heat may be transferred to the fluid through the boundaries of the system.

To visualize a system and energy transfer, think about the liquid in a metal pot on a stove. Energy is transferred from the hot stove surface to the liquid by conduction through the base of the pot, raising its temperature. Alternatively, fluid in a control volume could be heated with electrical heating tape surrounding a vessel to provide energy input—current flows through wires in the tape that have a substantial electrical resistance, so that electrical energy is converted to thermal energy that is conducted through the system walls. Or a coiled tube could be immersed in the fluid,

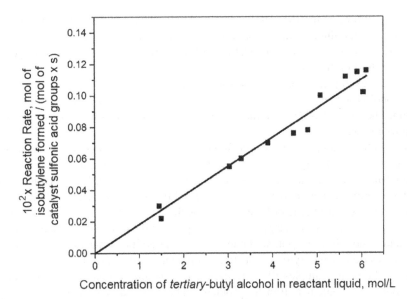

FIGURE 6.10 Data demonstrating that the reaction of *tertiary*-butyl alcohol catalyzed by particles of sulfonic acid resin is first-order in *tertiary*-butyl alcohol at the temperature of the azeotropic mixtures of water and *tertiary*-butyl alcohol in which particles of the catalyst were suspended. Do you understand why the line fitting the data must go through the origin? Can you figure out why the rate is normalized to the number of sulfonic acid groups? (This is a chemistry question.)

TABLE 6.4
Rates of Reaction of *tertiary*-Butyl Alcohol to Give Water and Isobutylene Determined with a Reactor Like That of Figure 6.9

Concentration of *tertiary*-Butyl Alcohol in Reactant Liquid (mol/L)	$10^2 \times$ Reaction Rate, mol of Isobutylene Formed/(mol of Catalyst Sulfonic Acid Groups $\times$ s)
1.45	0.030
1.50	0.022
3.04	0.055
3.30	0.060
3.90	0.070
4.48	0.076
4.80	0.078
5.08	0.100
5.65	0.112
5.91	0.115
6.04	0.102
6.11	0.116

with a hot fluid flowing through it and transferring heat by conduction through the tube walls to the fluid in the surrounding system; the temperature of the fluid in the coiled tube would fall from the inlet toward the outlet as it gave up heat to the surrounding fluid in the control volume.

In these examples, heat could be supplied or removed continuously, and we use the symbol Q for heat (energy) and dQ/dt for the rate of energy transfer in a batch system. The SI units of Q are Joules; rates of heat transfer in Joules per second give the power in watts.

As another example, consider a flowing fluid such as water or steam passing through a turbine. Energy is transferred, and there will be a number of terms in an energy balance equation. The mechanical energy driving the turbine is called work. Work is energy. As potential energy is converted into mechanical energy, the fluid is doing work. By causing the turbine to turn, the flowing fluid can be used to transform the energy into electrical energy. The moving parts in the flow system encounter friction, and so not all of the potential energy is converted into kinetic energy—some is lost—converted to energy that heats the equipment and the surroundings.

The following examples illustrate how the principle of conservation of energy is used to solve some heat transfer problems. The next example involves an antique flatiron: how it works and how an energy balance is used to analyze its performance.

Example 6.6 Heat Transfer and Performance of an Antique Flatiron

Problem statement: Explain how the flatiron shown in Figure 6.11 works by answering the following questions: How is the bottom surface of the flatiron heated? If the iron insert shown in the figure is heated in a wood fire, will it become hot enough to heat the flatiron to a temperature high enough to vaporize water in clothing that is being ironed? Will it be hot enough to smooth out wrinkles in cotton material? Will it possibly be too hot?

Approximate solution: The insert can be heated in a fire and then transferred into the slot in the flatiron. With the slot closed, the brass of the flatiron heats up as the iron heating element transfers heat to it. The hotter this element is initially, the hotter the flatiron becomes. We can look up temperatures of wood fires and find values ranging from about 770 K to much higher values (e.g., 1,300 K). Let us assume for simplicity that the heating element is brought to a temperature of 1,000 K before it is transferred into the flatiron. An energy balance gives an estimate of how hot the flatiron may become. This is an upper limit estimate because heat will be lost from the flatiron during and after heating.

Start with an energy balance: neglecting heat losses, all the heat transferred from the forged iron heating element goes to the flatiron: the heat capacity of the forged iron (C_{pFe}) is about 0.45 J/(g K) and that of the brass (C_{pB}) is about 0.38 J/(g K). If the terms M_B and M_{Fe} denote the masses of these two parts, and the starting temperature of the flatiron is about 25°C (298 K), and the initial and final temperatures are denoted with subscripts i and f, respectively, then the principle of conservation of energy implies that the heat given up by the insert is equal to the heat taken up by the flatiron:

$$M_{Fe} \times C_{pFe} \times \left(T_{Fei} - T_{Fef}\right) = M_B \times C_{pB} \times \left(T_{Bf} - T_{Bi}\right)$$

The respective masses are stated in the caption of Figure 6.11.

FIGURE 6.11   Antique flatiron made of brass (top) and piece of forged iron (bottom) that is heated and inserted to heat the flatiron. The flatiron has a mass of 1.82 kg, and the iron insert has a mass of 0.91 kg.

If we make the assumption that the temperature of the flatiron becomes equal to the temperature of the forged iron heating element without heat lost to the environment, then we set T_{Fef} equal to T_{Bf} and solve the equation to find the result that $T \approx 560$ K, or about 287°C. This is enough higher than the boiling point of water that it is evident that the flatiron initially would be able to vaporize water from materials such as wetted cotton. Suggested values of temperatures for ironing cotton and wool are about 200°C and about 150°C, respectively, which imply that the flatiron should be cooled before use with these fabrics. These values point to the appropriateness of the design of the flatiron and its heating system. The flatiron would cool during use, limiting the time of operation between insertions of heating elements, and so it would be helpful to have more than one heating element, with one being reheated while the other was being used.

HEAT CONDUCTION RATES AND THERMAL CONDUCTIVITY

Among the most easily understood examples of heat transfer and the use of energy balances are those involving heat conduction. Conduction of heat through a solid is associated with the motion of atoms and molecules in the solid. If we place the end of a metal spoon in a pot of boiling soup, the atoms in the spoon that are immersed in the liquid become hotter as heat is transferred from the soup to the spoon. The electrons at the immersed end of the spoon transfer heat along the spoon, so that it becomes hotter in our hand. There will be a temperature profile over the length of the spoon.

Some materials, such as metals, are good heat conductors, whereas others, which are often used as insulators, are not. Gases are poor heat conductors. Examples of good insulators are therefore materials that are porous solids—with air in open spaces in the solid (as in the straw or sawdust used to insulate blocks of ice transported to ice-box owners). Examples of porous metal oxides include magnesium oxide (magnesia), which in earlier times was combined with asbestos (which is avoided today, because it is toxic). Wood is also porous and a poor conductor of heat. A wooden spoon is therefore practical for stirring hot soup, whereas a metal spoon might be less appropriate.

We measure how well a material conducts heat with a fundamental physical property of the material, called the thermal conductivity. This is introduced in the following example.

Example 6.7 Heat Conduction in a Metal Rod

Problem statement: In a laboratory experiment, one end of a cylindrical metal rod with a diameter of 5.1 cm was heated by contact with a hot plate, and the other end was cooled with water flowing through a coil that was in contact with the cooled end of the rod and insulated so that the water flowing through it was heated by the rod and lost only a negligible amount of heat to the surroundings. Thermocouples were mounted in the rod to measure the temperature at various positions between the hot and cold ends. The experiment was carried out at steady state, and the rod was insulated to minimize the loss of heat from it, except at the cooled end. Data are presented in Table 6.5. Use the data to determine how the temperature depends on the distance between the ends of the rod.

Solution

A plot of the tabulated data is shown in Figure 6.12. The temperature decreases from the hot to the cold end, as expected, and the data fall near a straight line. Because the heat was being conducted along the rod at a steady state, the results suggest that the heat transfer rate was well approximated as a constant along the rod and proportional to the temperature gradient.

TABLE 6.5
Data Characterizing Steady-State Heat Transfer by Conduction through a Metal Rod Heated at One End and Cooled at the Other. Temperatures Were Measured with Thermocouples Embedded in the Rod

Distance from Hot End of Rod (cm)	Temperature (K)
8.4	414
16.9	403
25.1	395
33.5	390
41.9	378
50.3	373
58.7	363
67.0	356
75.4	348
83.8	339

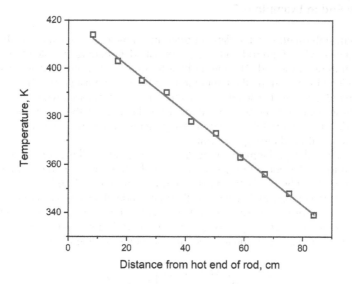

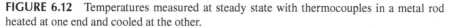

FIGURE 6.12 Temperatures measured at steady state with thermocouples in a metal rod heated at one end and cooled at the other.

Thus, we postulate on the basis of the linearity of the plot that the rate of heat transfer, dQ/dt, is described by the equation

$$\frac{dQ}{dt} = \text{const}\left(\frac{dT}{dz}\right) \tag{6.2}$$

where z is the axial distance from the hot end of the rod. This is a constitutive relationship. If the properties of the rod are the same throughout the rod (and if no heat is lost through the rod surface along its length because it is well insulated), we expect the rate of heat transfer to be proportional to the cross-sectional area of the rod, which we call A. Then we have an empirical constitutive relationship to model the system:

$$\frac{dQ}{dt} = kA\left(\frac{dT}{dz}\right) \tag{6.3}$$

Here k is the thermal conductivity (not to be confused with a reaction rate constant, which is represented with the same symbol). Thermal conductivity is a physical property that depends on the material through which the heat is conducted, with high values characterizing good conductors such as metals and low values characterizing poor conductors such as wood.

The data in this example illustrate a pattern that is observed generally, and Eq. (6.3) has a common name: *Fourier's law of heat conduction*. It is an essential constitutive relationship. An example showing how to determine values of thermal conductivity follows.

Example 6.8   Determining the Thermal Conductivity of the Rod in Example 6.7

Problem statement: Use the data reported in Example 6.7 to determine the thermal conductivity of the rod on the basis of data determining the rate of heat conduction through the rod. The rate of heat transfer through the insulated rod was determined by measuring the temperature of the cooling water into the coil cooling the rod to maintain a steady-state temperature profile, as shown in Figure 6.12. Use these data to determine the thermal conductivity of the rod if the cooling water flow rate was 0.398 g/s and the temperature of the cooling water increased by 53 K as it flowed through the coil.

We solve the problem by using an energy balance: the energy conducted through the rod is equal to the energy gained by the flowing cooling water: $dQ/dt = kA(dT/dz)$, where z is the axial distance along the rod and A is the cross-sectional area of the rod. Because the thermal conductivity k is well approximated as a constant, $dQ/dt = kA(\Delta T/\Delta z)$. This is equal to the rate of heat uptake by the cooling water, $(\Delta m/\Delta t)(C_{p\,water})(\Delta T_{water})$, where $\Delta m/\Delta t$ is the cooling water flow rate (mass/time), and ΔT_{water} is the temperature rise of the cooling water. Thus,

$$\frac{dQ}{dt} = kA\left(\frac{\Delta T}{\Delta z}\right) = \left(\frac{\Delta m}{\Delta t}\right)(C_{p\,water})(\Delta T_{water}) \tag{6.4}$$

Using data from Table 6.5, we have $\Delta T/\Delta z = (414 - 339)/(83.8 - 8.4) = 0.995$ K/cm. Using these values in Eq. (6.4) with the heat capacity of water stated above, we calculate that $k \approx 450$ W/(m × K), which is 450 J/(s × m × K).

Thermal conductivities of some materials are shown in Table 6.6. These data show a four-order-of-magnitude difference in thermal conductivity between a good insulator (brick) and a good conductor (copper). The data for copper show that the thermal conductivity depends on temperature, but, for many materials, data showing the temperature dependence are not readily available.

Looking at these data, what material might you suggest the rod in the preceding example was made of? Take account of the fact that errors in reported values of thermal conductivity of the order of 10% are not uncommon.

TABLE 6.6
Thermal Conductivities of Solids

Solid Material	Thermal Conductivity (W/(m K))
Insulating brick	0.15
Portland cement	0.29
Copper at 0°C	401
Copper at 327°C	383
Copper at 527°C	371
Glass	1.05
Magnesia insulation (85%)	0.07
Paraffin wax	0.25
Oak wood	0.17

Source: www.EngineeringToolbox.com, accessed May 26, 2022.

TABLE 6.7
Thermal Conductivities of the Wood Douglas Fir with Various Moisture Contents

Moisture Content of Wood (mass%)	Density of Wood (kg/m³)	Thermal Conductivity (W/(m × K))
0	385	0.086
8	385	0.093
12	385	0.097
16	385	0.10
0	545	0.097
8	545	0.109
12	545	0.113
16	545	0.118

Source: F. B. Rowley and A. B. Algren, Bull. 12, Engineering Experiment Station, University of Minnesota, 1937.

Example 6.9 Thermal Conductivity of a Porous Material: How Water Content Affects Wood as a Heat Conductor

Problem statement: The data shown in Table 6.7 indicate how the thermal conductivity of Douglas fir depends on its moisture content and density. Summarize the trends in the data and explain them.

Solution

Qualitatively, two patterns are clear from the data: as the moisture content of the wood increases, the thermal conductivity increases. And as the density increases, the thermal conductivity increases. We infer that as water is added to the wood, more of the open (pore) spaces are filled with water, and liquid water is a better conductor of heat than air, so that the thermal conductivity increases. Increasing the density of the wood, separately from the water content, also leads to a higher thermal conductivity, indicating that as the wood molecules (these are long polymeric molecules) become closer to each other, on average, the energy transfer between them becomes faster. Do you think there might be some indications in these data that could help predict the course of a forest fire? What thoughts do you have about the aging and storage of lumber or logs before they are cut into lumber?

INSULATION AND ENERGY CONSERVATION

Insulating materials are widely used to minimize energy losses from many kinds of apparatus as well as buildings. Insulation helps retain energy in a building when it is cold outside and helps keep a building cool when it is hot outside. Building insulation is often made of fiberglass and is available as easily installed blankets, or it can be sprayed into spaces between walls. It needs to fill these spaces, because heat transfer by convection can be significant in non-filled spaces like corners.

Some forms of insulation must withstand high temperatures and challenging chemical environments, such as in furnaces. These *refractory* materials are usually porous, offering low thermal conductivities combined with robustness.

Fourier's law combined with energy balances provides the basis for analysis of heat transfer by conduction through walls, including composite walls—those having layers of separate materials, which are commonly used.

The following is a development of the energy balance equations for steady-state conduction through a composite plane wall consisting of three separate materials constituting the wall of a furnace. For simplicity, we approximate the walls in the system to be an infinite flat plane to avoid complications of curvature of the wall and of edges and corners in the wall. We assume that the wall separates the hot inside of a furnace, where fuel is burned, from the cooler outside.

At steady state, the rate of heat conduction through an area A of the wall is the same for each section (layer), and, by Fourier's law, is the following:

$$dQ/dt = \frac{k_{12}A(T_1 - T_2)}{\Delta z_{12}}$$ (6.5)

for the first layer (Δz_{12} is the thickness of the first layer, the one in direct contact with the burning fuel). Considering several layers with different compositions, we have started by writing Eq. (6.5) for the first layer, with k_{12} being the thermal conductivity of the material in that layer. The first layer might be firebrick, for example, chosen to withstand the atmosphere of the burning fuel.

For the second layer, which we might designate to be insulating brick, the corresponding equation is the following, where k_{23} is the thermal conductivity of the insulating brick:

$$dQ/dt = \frac{k_{23}A(T_2 - T_3)}{\Delta z_{23}}$$ (6.6)

For the third layer, the corresponding equation is the following, where k_{34} is the thermal conductivity of the third kind of brick:

$$dQ/dt = \frac{k_{34}A(T_3 - T_4)}{\Delta z_{34}} \qquad (6.7)$$

Doing some algebra, we find

$$T_1 - T_4 = \frac{dQ/dt}{A}\left[\frac{\Delta z_{12}}{k_{12}} + \frac{\Delta z_{23}}{k_{23}} + \frac{\Delta z_{34}}{k_{34}}\right] \qquad (6.8)$$

This result is easily generalized and represented simply as

$$\frac{dQ/dt}{A} = \frac{\Sigma(\Delta T)}{\Sigma\left(\dfrac{\Delta z}{k}\right)} \qquad (6.9)$$

where each value of k in the summation represents a particular material.

Example 6.10 Heat Transfer through the Walls of a Furnace

Problem statement: A firewall of a combustion furnace is installed to minimize heat losses to the surroundings. Consider a furnace with combustion occurring at steady state that has a wall consisting of one kind of brick, with a thickness of 0.25 m and a thermal conductivity of 1.2 watt/(m × K); with a layer of insulating brick inside it, having a thickness of 0.125 m and a thermal conductivity of 0.15 watt/(m × K); and a third kind of brick in a layer inside the insulating brick, having a thickness of 0.20 m and a thermal conductivity of 0.70 watt/(m × K). Assume that the inside wall temperature is 1,100 K (i.e., assume that the burning mixture is well mixed) and the outside wall temperature is 340 K.

Find the rate of heat transfer per unit area of the wall and find the temperatures at the interfaces between the walls that consist of different components. Ignore heat transfer resistance in the mortar connecting the three brick layers of the wall. Simplify matters by assuming that the wall can be considered to be an infinite flat plane.

Solution

Using Eq. (6.9) and the data provided, we find

$$\frac{dQ/dt}{A} = \frac{(1100 - 340)}{\left(\dfrac{0.25}{1.2} + \dfrac{0.125}{0.15} + \dfrac{0.20}{0.70}\right)} = 573 \text{ watt/m}^2$$

To find the temperatures between the layers of brick, we start with the following rearranged form of Eq. (6.5) and the value of $\dfrac{dQ/dt}{A}$:

$$\frac{dQ/dt}{A} = \frac{k_{12}(T_2 - T_1)/\Delta z_{12}}{\Delta z_{12}} = 573 \text{ watt/m}^2 \qquad (6.10)$$

determining a value of T_2 of 483 K.

Similarly, we determine $T_3 = 961$ K. These results show the effectiveness of the insulating brick as a good insulator: $T_2 - T_1$ is only 143 K, whereas $T_3 - T_2$ is 478 K.

TEMPERATURE CONTROL AND TEMPERATURE CONTROLLERS

Many applications require control of temperatures. Devices that provide temperature control are common (including thermostats in many buildings), and they include a temperature sensor such as a thermocouple. The thermocouple provides an electrical signal (voltage) from the place where the temperature is to be controlled. Another signal corresponds to a set point, the desired temperature, chosen by the user. Such controllers today are electronic.

A common goal is to control the temperature of a system. Consider, for example, a fluid in a tank to be the system. The fluid may be mixed, which reduces the variations from place to place in the system. The temperature-control device provides heat input or cooling, guided by the difference between the measured temperature and the set temperature. A fluid in a tank, for example, could be heated by a hotter fluid flowing through a coil of pipe immersed in the tank, or cooled by a colder fluid flowing through the coil. (This coil or pipe is a simple heat exchanger. Many heat exchangers are more intricate in design, consisting of multiple parallel tubes with flowing coolant (or the reverse) surrounded by a shell with another flowing fluid to be cooled (or heated); this is called a shell and tube heat exchanger.)

In an on-off controller, the flow of the heating or cooling fluid is either on or off. Thus, when there is a difference between the recorded system temperature and the set temperature, there is flow of fluid in the coil; otherwise, there is no flow. The temperature that is being raised or lowered may overshoot under the influence of the controller, and so the temperature may oscillate before coming to a steady state.

A more sophisticated and more effective design (which usually drives the system to the set temperature faster than an off-on controller) is a proportional controller: the rate of heat input or withdrawal (e.g., controlled by the flow rate of the heating or cooling fluid (heat transfer fluid) in the coil) is proportional to the difference between the system temperature and the set temperature. The signal from the controller would determine how far open the valve regulating the flow of the heating or cooling fluid would be.

Temperature controllers may be programmed for optimum operation by responding to the signals in ways that are much more complex than what we have described; for example, there may be a nonlinear dependence of the rate of flow of heating or cooling fluid on the difference between the system and set temperatures. Analysis of these devices is beyond the scope of this book and requires mathematics that we have not learned yet.

The following example illustrates how a temperature controller is used in the design of a system for measuring the rate of a chemical reaction.

Example 6.11 Measuring Rates of a Reaction Catalyzed by Solid Particles

Problem statement: Many chemical reactions are catalyzed by solids that are used in the form of particles that are placed in reactors. The reactors are often tubes that are well approximated as piston-flow reactors. A stream of reactant gas or liquid flows into one end of the tube, comes in contact with the catalyst, and undergoes conversion. A product stream, including unconverted reactants, then usually flows out of the tube to purification devices, and unconverted reactants may be recycled to the feed stream. The catalyst particles remain in the reactor, which is called a packed bed (or a fixed bed). These particles are usually porous, with high internal surface areas, even as high as hundreds of square meters per gram. The reactions take place on the internal catalyst surfaces, and so high surface areas per unit volume of reactor lead to high rates of reaction per unit volume of reactor and therefore efficient use of reactor volume. Often rates of the catalytic reactions are determined by analysis of the product stream, but some reactions allow simpler determinations of these rates. We consider an example of one of these here.

The reaction of formic acid to give water and carbon monoxide,

$$HCOOH \rightarrow CO + H_2O \tag{6.11}$$

is catalyzed by a number of solid acids (as well as liquid acids). One such catalyst has SO_3H groups (sulfonic acid groups) on its internal surface, and these have chemical properties similar to those of sulfuric acid, H_2SO_4. When gas-phase formic acid comes in contact with these SO_3H groups, it reacts to form water and carbon monoxide.

Design a reactor to determine rates of this reaction (without relying on chemical analysis of the products) at various temperatures. Explain how the reactor and temperature control work.

Solution

A simple design takes advantage of the fact that CO has a boiling temperature of 81.6 K at atmospheric pressure (normal boiling point) that is much lower than that of formic acid (which has a normal boiling point of 374 K) or of water (which has a normal boiling point of 373 K). Thus, we see an opportunity to easily separate the CO from the other components by using a water-cooled condenser, with a cooling water temperature of about room temperature, for example.

A suggested design is shown in Figure 6.13. How does it work?

Particles of catalyst are placed in the tube at the left, held in place by inert glass wool on a porous frit. Thus, the catalyst particles are in a fixed bed in the reactor tube, which is approximated as a piston-flow reactor. The catalyst particles are mixed with inert particles of glass to increase the rate of heat transfer to and from the catalytic particles and make the fixed-bed temperature nearly constant. The tube is heated by the electrical resistance of a wire wrapped around it. Voltage is applied to the wire, and the higher the voltage, the hotter the wire (we see this from Ohm's law). An electronic temperature controller, with input from a thermocouple placed in the well inside the catalyst bed, generates the signal for the temperature controller, determining the energy input to keep the fixed bed at the desired temperature. The temperature is varied by changing the controller set point to allow measurements of reaction rates at each of a set of various temperatures.

Liquid formic acid is placed in the flask at the bottom of the apparatus, which is heated electrically. The formic acid boils, and its vapor flows up through the temperature-controlled fixed bed of catalyst particles, where the reaction to give water and carbon monoxide takes place. The more catalyst there is in the fixed-bed reactor, the higher the conversion of the formic acid that flows through it. The vapor stream exiting the fixed-bed reactor flows to a water-cooled condenser, where the water and unconverted formic acid are condensed so that they can be returned to the flask, but the volatile carbon monoxide is not condensed and flows out of the device and to the gas flow meter (a soap-film flow meter) for determination of the flow rate, which determines the rate of the catalytic reaction.

Carbon monoxide is toxic, and so it must be vented safely, for example, into a laboratory fume hood that exhausts the carbon monoxide to the outside air.

Example 6.12 Determining the Mass of Solid Catalyst in a Reactor for Formic Acid Dehydration

Problem statement: Apparatus like that shown in Figure 6.13 was used to determine rates of formic acid conversion into water and carbon monoxide catalyzed by a polymer incorporating sulfonic acid groups (SO_3H groups). The rate was found to be 8×10^{-2} (mol of formic acid converted)/(mol of SO_3H groups $\times$ min) at 410 K and 1.0 atm of formic acid. Figure out how to make a measurement to check this result using the apparatus shown in Example 6.11. Thus, if the mass of catalyst in the tube is 1.0 g and the catalyst contains 5×10^{-3} mol of SO_3H groups/g, what is the expected volume flow rate of CO that would be measured with a soap-film flow meter just after the flow of formic acid vapor started? Assuming that the soap-film flow meter works at 298 K and 1.0 atm, would this be an appropriate rate for accurate measurements with the soap-film flow meter?

What restrictions would we need to place on the conversion in the fixed bed?

The catalytic reaction has been found to be inhibited by water. What are the consequences of this inhibition for the experiment to be done with the apparatus shown in Figure 6.13?

Solution

The stoichiometry of the reaction shows that one mol of formic acid reacts to give one mol of water and one mol of carbon monoxide, so that the rate of consumption of formic acid in units of mol/time is equal to the rate of formation of carbon monoxide in mol/time. This reaction rate corresponds to the rate of flow of carbon monoxide in the off-gas stream. We need to calculate this reaction rate as a volume flow rate at 298 K and 1.0 atm, because these are conditions of the gas flowing through the soap-film flow meter. We calculate that from the given information, as follows: the stated rate of reaction is 8×10^{-2} (mol of carbon monoxide formed)/(mol of SO_3H groups $\times$ min), and the number of mols of SO_3H groups is 5×10^{-3} (the number of mols in 1.0 g, which is the given mass of catalyst), and, so, using this number of mols of SO_3H groups in 1.0 g of catalyst, we calculate the rate of carbon monoxide formation to be 8×10^{-2} (mol of carbon monoxide)/(mol of SO_3H groups $\times$ min) $\times (5 \times 10^{-3}$ mol of SO_3H groups$) = 4.0 \times 10^{-4}$ (mol of CO)/min.

Because the temperature and pressure are close to ambient, and the ideal gas law accurately represents the low-molecular-weight carbon monoxide under ambient conditions, we use the ideal gas law to convert the reaction rate to a volume flow rate, V/t, as follows: $V/t = NRT/tP = [(4.0 \times 10^{-4}$ mol/min$^{-1})(0.08206$ L atm mol^{-1} K$^{-1})(298$ K$)/1.0$ atm$] = 9.8 \times 10^{-3}$ L/min^{-1}.

This is 9.8 mL/min, indeed a convenient flow rate to measure with a standard soap-film flow meter like the one shown in Chapter 2, in which the measurement volume can easily be set to be in the range of 1–20 mL. Thus, the times for flow of carbon monoxide in the set flow meter volume are of the order of 1 min, which can be measured with good accuracy (of the order of ±1%) with a stopwatch.

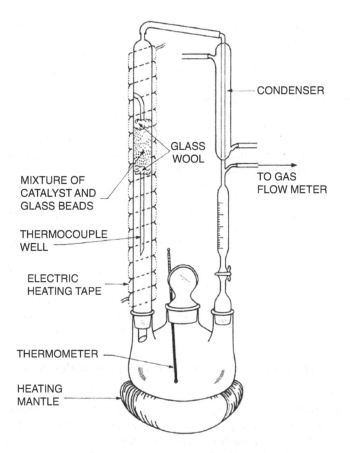

FIGURE 6.13   Reactor with temperature-controlled fixed bed of catalyst particles designed for measurement of rates of reaction of formic acid conversion to carbon monoxide and water. The fixed-bed reactor is approximated as a piston-flow reactor, and its temperature is regulated with a controller. The heat generated by the metal wire in the heating element is determined by the electrical current and the resistance of the wire. The controller senses the temperature in the fixed bed with a thermocouple in the thermocouple well inside the fixed bed, and the current in the heating tape is determined by the difference between the temperature inside the fixed bed and the set temperature. The unconverted formic acid and water vapor flowing upward from the reactor are condensed and may be returned to the liquid reservoir (which is heated electrically with a heating mantle), from which they are vaporized to flow up again through the fixed-bed reactor. The carbon monoxide formed in the reaction has such a low boiling point that it is not condensed and instead flows to the gas flow meter, where its flow rate (the reaction rate) is determined. Figure reproduced with permission from B. C. Gates and J. D. Sherman, Experiments in Heterogeneous Catalysis: Kinetics of Alcohol Dehydration Reactions, *Chem. Eng. Educ.*, **124**, summer 1975.

A limitation to consider in the above statement is the following: we have assumed that the rate is the same throughout the fixed bed of catalyst, but in a plug-flow reactor, the conversion will change between the inlet and the outlet, because the concentration of reactant (and products) will change between the inlet and the outlet. Only if the conversion is low will the concentrations of reactants and products be approximately the same at the inlet and the outlet of the fixed bed and the rate nearly the same throughout the fixed bed. When that criterion is met, we refer to the reactor as a differential reactor—one in which the rate changes negligibly between inlet and outlet and can be approximated as a constant. Differential reactors are often used to measure reaction rates.

This reaction rate is the rate that would be observed once the temperatures had achieved steady state and the flow rate of formic acid had become nearly constant through the fixed bed of catalyst particles. But because the reaction produces water, and water is accumulated in the vessel with the boiling formic acid, the composition of the vapor flowing up from that vessel will contain more and more water as the experiment proceeds if the liquid accumulating in the lower right of Figure 6.13 is allowed to flow down to the reactor (if the valve is opened). Thus, the rate of the reaction will decline as the experiment proceeds, because of the inhibition of reaction by water. To deal with this complexity, we would collect volume flow rate data for carbon monoxide as a function of time of operation and extrapolate the values to time = zero—this would provide a good measure of the rate of the reaction in the limiting case for which no water was present. Mass balance calculations determined from the integrated reaction rate data determine the composition of the liquid as a function of time of operation and allow measurements of rate for various compositions of the reactant gas stream, but remember that the composition of the gas phase formed from the liquid mixture is different from that of the liquid because formic acid and water have different normal boiling points.

CALORIMETRY, HEATS OF MIXING, AND HEATS OF CHEMICAL REACTION

We have seen that phase changes (melting/freezing and vaporization/condensation) are characterized by energy changes. Other physical and chemical changes are also generally associated with energy changes. For example, dissolving a solid in a liquid or mixing one liquid with another is associated with a heat of solution or a heat of mixing. The values can be positive or negative: heat may be either given off or taken up during these changes. Chemical reactions are characterized by heats of reaction, which can also be positive or negative.

How are these energy changes measured? A standard approach is to use an apparatus called a calorimeter. A calorimeter is designed so that almost all the heat characterizing the change in the sample is transferred to or from a reference fluid or solid. The apparatus works by measurement of the temperature change of the reference fluid or solid. Combining knowledge of this temperature change with knowledge of the heat capacity of the reference fluid or solid—combined with an energy balance—tells us how much the energy change is. Good heat transfer between the changing sample and the reference fluid or solid is needed, and we have to recognize that not all the heat transferred to or from the sample may go to the reference fluid or solid. Some heat will also be transferred to or from the calorimeter. Good insulation is needed to minimize heat transfer to or from the surroundings. Calorimeters are designed to minimize these heat losses or gains. There are many designs; some calorimeters are batch and some are flow systems, and those designed for determining heats of reaction are of course reactors.

In summary, the gist of the operation of a calorimeter is that almost all of the heat generated or taken up in a process such as dissolution, mixing, or chemical reaction is transferred to (or from) the sample being investigated and correspondingly from (or to) a reference liquid or solid of known mass, and the amount of heat transferred is measured by the temperature change, the known mass, and the heat capacity of the reference fluid or solid. Provided that the insulation is good and the design is effective, almost all of the heat is transferred to or from the reference fluid, and then a simple energy balance provides the basis for data analysis. The reference fluid is commonly water. A key point is that measurements of temperature determine energy changes.

Example 6.13 Determining Heat of a Chemical Reaction with a Bomb Calorimeter

Problem statement: The apparatus called a bomb calorimeter was used for burning 3.12 g of glucose, $C_6H_{12}O_6$, in the presence of a stoichiometric excess of oxygen, so that the conversion of the glucose was complete. The temperature of the calorimeter increased from 297.0 K to 308.8 K, for a change of 11.8 K. The calorimeter contained 0.775 kg of water, and it had a heat capacity, presumably measured in calibration experiments, of 0.893 kJ/K. Determine how much heat was produced by the combustion of the glucose and express the result in units of J/mol of glucose converted. (Source of data: OpenStax, referenced June 30, 2022; statement of problem modified from that in OpenStax.)

Solution

The combustion reaction produced heat that was transferred to the water and to the bomb calorimeter (plus small amounts of heat absorbed by the reaction products and the unreacted excess oxygen that was present initially, and these are assumed to be negligible in the calculation). The energy balance is a statement that the heat given up by the chemical reaction (we call it $-Q_{reaction}$—by convention the heat of reaction has a negative value when the reaction gives off heat—is exothermic) is the heat taken up by the water and the calorimeter, provided that the heat lost to the surroundings is negligible by comparison with these terms. Therefore, we write the following energy balance equation, using the value of the heat capacity of water taken from Table 6.2:

$$-Q_{reaction} = Q_{water} + Q_{bomb} = m_{water}\left(C_p\right)_{water} (\Delta T)_{water} + m_{bomb}\left(C_p\right)_{bomb} (\Delta T)_{bomb} \qquad (6.12)$$

$$-Q_{reaction} = \left(0.775 \text{ kg} \times 4.19 \text{ kJ kg}^{-1} \text{K}^{-1} \times 11.8 \text{ K}\right) + \left(0.893 \text{ kJ K}^{-1} \times 11.8 \text{ K}\right) = 48.9 \text{ kJ} \qquad (6.13)$$

Using the molecular mass ("molecular weight") of glucose (180 g/mol), we calculate the number of mols of glucose (0.0173 mol) and thus the heat of reaction of 2.8×10^3 kJ/mol of glucose converted. Notice how large this value is by comparison with the heats of solution that are mentioned below.

Calorimeters are used to quantify energy changes associated with many physical processes, including heats of solution, which are those occurring when two or more liquids or a liquid and a solid or a liquid and a gas are mixed with each other. Mixing experiments in calorimeters have shown, for example, that when sodium chloride and water are mixed with each other, the temperature changes only little—the heat of mixing is roughly 4 kJ/(mol of NaCl) over a wide range of NaCl concentrations, and heat is given up by the mixture—the temperature decreases when the two components are mixed.

Other combinations are characterized by orders of magnitude greater heats of mixing. For example, when concentrated nitric acid in water is made, the heat of solution is about 38 kJ/mol of nitric acid, and the process gives off heat—the solution becomes hotter upon mixing of the two components. When sulfuric acid is used instead of nitric acid, almost twice as much heat is given off per mol of acid when it is mixed with water to give a highly concentrated solution of the acid. Consequently, a person working in a laboratory needs to be safety conscious when mixing these acids and water.

With heat of solution data and heat capacity data, one can determine the temperatures and compositions of solutions that result when multiple components are mixed, presuming that they do not undergo a chemical reaction. The next example provides an illustration of how such results are determined—it is significant as an illustration of how we use an energy balance combined with a mass balance.

Example 6.14 Heat of Mixing of Liquids

Problem statement: Consider the mixing of an aqueous solution of potassium hydroxide (KOH) with water in a tank that is so well insulated that heat losses/gains can be neglected. The mass of water to be added to the tank at 298 K is 800 kg, and the mass of an aqueous solution containing 5.0 mass% KOH at 298 K to be added is 400 kg. Estimate the concentration of the resulting solution in mols of KOH per liter, and estimate how much heat would have to be removed from the tank for the resulting solution temperature to be 298 K. The heat of solution of the KOH solution in water under these conditions is 12.4 kcal/(mol of KOH)—the process gives off heat (it is exothermic). If the heat were not removed but instead used to heat the solution in the tank, what would the solution temperature be if the tank were perfectly insulated?

Solution

We estimate the concentration of KOH in the resulting solution by doing a mass balance calculation and looking up the density of the solution. We leave out the details here (because we have done similar calculations before) and find that the solution contains 344 mols of KOH and, with the corresponding mass fraction of KOH, a density that we look up and find to be 1.02 kg/L, giving a volume of 1180 L and a KOH concentration of 0.29 mol/L. Given that the heat released is 12.4 kcal/(mol of KOH), we use the number of mols of KOH to find that the heat released is 12.4 kcal/mol × 344 mol = 4.3 × 10^3 kcal, which is (4.3 × 10^3 kcal)(4.18 kJ/kcal) = 1.8 × 10^4 kJ. To estimate the temperature rise in the perfectly insulated tank, we approximate the heat capacity of the solution as almost the same as that of water, 4.2 kJ/(kg × K) (because the KOH solution is dilute) and ignore the temperature dependence of the heat capacity, and do an energy balance:

$$Q = mC_p\Delta T = mC_p(T - 298) \qquad (6.14)$$

$$1.8 \times 10^4 \, \text{kJ} = (1200 \, \text{kg}) \big[4.2 \, \text{kJ/(kg} \times \text{K)} \big] (T - 298 \, \text{K}) \qquad (6.15)$$

Thus, we find that $T = 302$ K. The result is that the temperature increases by about 4 K because of the exothermicity of the dissolution of the KOH solution in water.

This example shows that a temperature rise of only a few degrees characterizes the mixing of a dilute solution of KOH with an excess of water. However, sometimes, when solutions of other components are mixed, the temperature rises can be orders of magnitude greater than this.

For example, in some chemical processes, highly concentrated sulfuric acid is used as a reactant or as a catalyst. A solution of 95 mass% H_2SO_4 in water has an acid concentration of about 18 mol/L. Adding this to an equal volume of water can lead to a temperature rise of the order of 100 K. The combination of the acid and water involves significant chemical change—when sulfuric acid is combined with water, it undergoes highly exothermic reactions, primarily including proton transfer to make H_3O^+ and HSO_4^-. Consequently, safety considerations are a dominant concern. When water is added to the concentrated acid without good mixing, the temperature can rise quickly where the two components come in contact with each other, causing boiling/sputtering of the solution and endangering an experimenter. It is much safer to add the acid to the water—slowly—and with good mixing. Then the denser acid may sink to the bottom of the container before forming a solution with the water and be less prone to boil and sputter.

CALORIMETRY FOR DETERMINING RATES OF CHEMICAL REACTIONS

Chemical reactions give off or require energy, and the rate of heat generation or uptake by a reaction is proportional to the rate of the reaction. This simple relationship means that measurements of rates of heat generation or uptake can be used to determine reaction rates, and so rates of reactions can be determined simply with calorimeters. Calorimeters that are used as reactors are normally batch reactors, and they are similar to the calorimeters described above. The rate of heat transferred from (or to) the reactor is the following (presuming no volume change on reaction, no heat losses from the calorimeter, and a constant temperature of the reaction vessel):

$$\frac{dQ}{dt} = Q_{rxn} V r \qquad (6.16)$$

where we are calling the heat of reaction Q_{rxn} in dimensions of energy per mol (e.g., J/mol), V is the volume of reactant, and r is the reaction rate. (With this notation, Q and Q_{rxn} have different dimensions.)

Calorimeters that are equipped to be reaction calorimeters are efficient tools for measuring reaction rates, offering the advantages of essentially continuous monitoring of rates without the need for sampling and chemical analysis of the reaction and product mixtures to determine conversions. The same apparatus can be used to determine both reaction rates and heats of reaction.

Chemical engineers have used reaction calorimeters to determine kinetics of reactions that are important in the manufacture of pharmaceuticals, using the results for reactor scale-up and process design. For example, an amination reaction, that is, one that involves reaction with a compound called an amine (exemplified by RNH_2, where R represents an alkyl group such as a methyl group, CH_3; a specific compound that is an amine is $H_3C(CH_2)_4NH_2$). The reaction takes place as follows (but this statement is simplified, because more than one amine was present reacting with the bromine-containing aromatic compound at the left, and a catalyst—not shown—was also present):

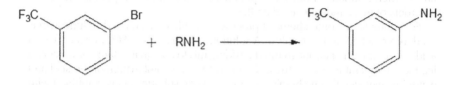

The full stoichiometry of the reaction is complicated and not shown here; bromine ends up in the product sodium bromide, NaBr. In the laboratory experiments described here, the reactor volume was 6.0 mL, and a vessel containing 4.5 mL of toluene was placed in the reference chamber of the calorimeter where its temperature was monitored. Each vessel was mixed with a stir bar, magnetically driven, to ensure essentially perfect mixing. The temperature of the toluene was controlled to be nearly constant, and the rates of heat transfer to keep this temperature constant were measured, determining reaction rates.

Data were recorded with such a high frequency that they nearly fall on a line (Figure 6.14). To repeat, the data include rates of heat transfer from the reactor that determine reaction rates (the black curve), by Eq. (6.16), and also conversions (the gray curve), which were determined by integration of the rate data. Can you explain how the gray curve was determined from the black curve (this is a calculus question)?

An Industrial Process with Reactors Designed for Heat Balance: Endothermic Catalytic Cracking in a Reactor That Is Coupled with Exothermic Coke Burning in a Well-Mixed Regenerator Reactor

Many industrial reactions are strongly exothermic, and many are strongly endothermic. Consequently, the reactors must be designed for rapid heat transfer. If the reactions are catalyzed by solids, as is typical, heat must be transferred through the solid, which is usually present in the form of particles. The thermal conductivity of a fixed bed of particles is markedly less than that of an individual particle itself, because heat transfer in the spaces between the particles—which are usually filled with reacting gases—is slower. If a solid-catalyzed reaction is strongly endothermic and is started at a high rate, the rate will decline, often quickly, if heat is not supplied to the catalyst fast enough, because the catalyst will be cooled by the endothermic reaction, and the rate depends exponentially (by the Arrhenius equation) on the absolute temperature. On the other hand, if a solid-catalyzed reaction starts at an appropriate rate and the reaction is exothermic, the heat given off by the reaction—if it is not removed—will raise the temperature of the catalyst (and the reactor), and the reaction rate will increase exponentially—sometimes with catastrophic

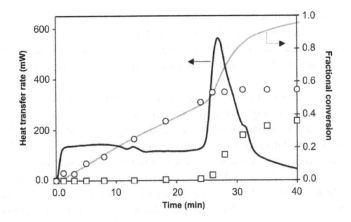

FIGURE 6.14   Data characterizing aromatic amination reaction shown above taking place in a calorimeter at 363 K. The circles and squares represent fractional conversion of the reactant 3-bromobenzotrifluoride to the products, as determined by chemical analysis by gas chromatography of samples taken from the reactor. The black line shows the rate of heat transfer from the reactor (the heat generated by the reaction), which determines rates of reaction. The fractional conversion shown on the plot with the gray line was determined by integration of the heat transfer rate data. Reproduced with permission from A. C. Ferretti, J. S. Matthew, I. Ashworth, M. Purdy, C. Brennan, and D. G. Blackmond, Mechanistic Inferences Derived from Competitive Catalytic Reactions: Pd(binap)-Catalyzed Amination of Aryl Halides, *Adv. Synth. Catal.*, **2008**, *350*, 1007–1012. O. P. Schmidt and D. G. Blackmond, Temperature-Scanning Reaction Protocol Offers Insight Into Activation Parameters in the Buchwald-Hartwig Pd-Catalyzed Amination of Aryl Halides, *ACS Catal.*, **2020**, *10*, 8926–8932.

consequences, such as an explosion or vaporization or melting of the catalyst or melting of the reactor.

Fluidized beds are often used for strongly exothermic reactions, such as oxidations, because the moving particles cause rapid mixing of the fluid–solid mixture, and this rapid mixing greatly increases the rate of heat transfer in the reactor—this is heat transfer primarily by convection.

We illustrate some of these points and the importance of managing energy on a massive scale by considering catalytic cracking, a petroleum-refining process for converting hydrocarbon molecules in oil into smaller molecules, those that have boiling ranges appropriate for gasoline. Catalytic cracking is one of the largest-scale industrial processes.

The cracking reactions of the many and various hydrocarbon molecules in petroleum (e.g., *n*-dodecane) are highly endothermic, requiring lots of heat input into the reactor if high conversions are desired (and they are). Accompanying the cracking reactions are reactions that form high-molecular-weight material called coke (a complex material with an average composition of approximately CH), which deposits on the catalyst particles and fills their pores, limiting the access of reactant molecules by diffusion through the pores of the catalyst particles. A consequence of the coke formation and deposition is that the catalyst loses activity—the rate of the catalytic reaction becomes less as the catalyst is deactivated. In a catalytic cracker, this deactivation takes place in a matter of just a few seconds.

A design that accommodates this rapid deactivation is one that involves two reactors working in tandem. The design allows reactivation of the catalyst in a reactor separate from the cracking reactor and at the same time provides heat to the cracking reactor to keep the temperature high enough for high rates of cracking and high conversions.

The design is shown in Figure 6.15. What this shows, at the right, is a cracking reactor that operates at high temperatures and in which the catalyst particles are entrained—carried from the bottom toward the top of the reactor by the flowing reactant gas stream. This reactor is called a riser reactor—it is a fluidized-bed reactor that is an entrained-bed reactor. The gas and particles reside in this reactor for only several seconds (during which the catalyst loses a substantial fraction of its activity as the temperature falls). At the top of the reactor is a separations device—a cyclone—where the gas-phase products are separated from the catalyst particles. The catalyst particles then move to another reactor, called a regenerator, where they are exposed to air and where the coke burns, giving off carbon dioxide, steam, and lots of heat. This reactor is a fluidized bed, very well mixed, so that the catalyst particles that leave the reactor are hot—heated by the exothermic coke combustion reaction (Figure 6.15). The hot catalyst particles are recycled to the riser reactor. Thus, by carrying heat to the riser reactor, the catalyst particles provide the energy needed for the endothermic cracking reactions, and complete a cycle.

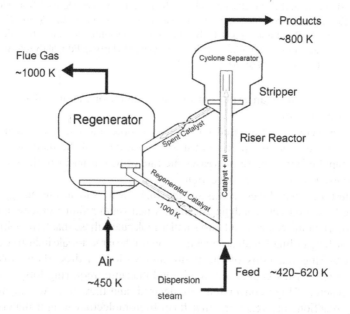

FIGURE 6.15 Process flow diagram of a catalytic cracker with a regenerator, a system of reactors for converting petroleum fractions into lower-molecular-weight petroleum fractions including those in the gasoline-boiling range. The design illustrates an exquisite heat balance: heat generated in the coke combustion in the regenerator is transferred to the catalyst particles, and when they are recycled to the riser reactor, they provide heat for the endothermic cracking reactions. The catalyst particles in the riser reactor are cooled by the endothermic cracking reactions as they rise.

There is energy balance between the two reactors. The design demonstrates an exquisite heat balance. The heat needed for the cracking reaction in the riser reactor is supplied in the regenerator reactor by burning of the undesired product, coke, that forms along with the desired cracking products.

Catalytic crackers are marvels of chemical engineering design. They demonstrate many of the principles of chemical engineering.

AN INDUSTRIAL PROCESS THAT COMBINES BIOMASS CONVERSION WITH PETROLEUM CONVERSION

When biomass, which consists primarily of compounds made of carbon, hydrogen, and oxygen, is converted into chemicals and into fuels, and when these are burned to generate energy, they make carbon dioxide, which is converted (by photosynthesis) back to biomass such as trees—the carbon is recycled. With a strategy of replacing petroleum and other fossil resources with biomass, we can minimize loading the atmosphere with the greenhouse gas carbon dioxide.

The process design for petroleum cracking described in the preceding section offers lots of flexibility including opportunities for converting biomass, along with petroleum. An energy transition from hydrocarbons to renewables such as biomass (e.g., wood chips, fast-growing grasses, algae) is well underway, involving many chemical engineers. A new multizone catalytic cracking process (Figure 6.16) converts entire petroleum feedstocks, with zones in the riser reactor controlled for operation under various conditions to control the product distribution. The design is flexible to accommodate biomass-derived feed streams. Various feeds are introduced at various heights along the riser, and the riser reactor diameter increases with each introduction of feedstock to allow for the increasing flow rate. The feed introduced at the inlet of the reactor includes difficult-to-crack components from a petroleum refinery, with more reactive crude oil introduced higher up in the reactor, and the oxygen-containing biomass-derived components (oxygenates), which are more reactive, being introduced still higher up in the reactor (Figure 6.16).

Again, the design allows continuous catalyst regeneration and a heat balance. It provides technology that is making a transition to a more nearly circular economy and demonstrating the importance of chemical engineering in that transition.

RECAP AND REVIEW QUESTIONS

The principle of conservation of energy is a foundation for analysis that is as basic and essential as the principle of conservation of mass, and it is used as the starting point for addressing many energy transfer phenomena. We have illustrated it here to account for heat conduction by using a simple constitutive relationship (Fourier's law) and a relationship defining the heat capacity. Heat is transferred by conduction, convection, and radiation, and we emphasize conduction here to keep matters simple mathematically. Energy is central to this chapter, and it is helpful to consider various forms of energy and how changes in internal energy of gases, liquids, and solids are associated with phase changes and chemical reactions. These energies need to be accounted for in many designs. Can you see how this chapter builds on ideas

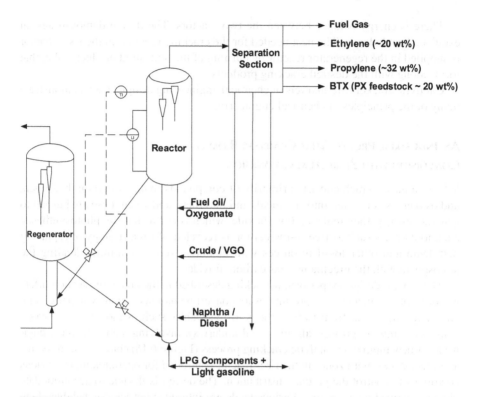

FIGURE 6.16   Process flow diagram of a modified catalytic cracker with a regenerator, a system of reactors for converting various petroleum fractions and biomass-derived feedstocks (oxygenates) into valuable chemicals and fuels. Terminology: LPG, light petroleum gas; naphtha, hydrocarbons boiling in the gasoline range; VGO, vacuum gas oil, a fraction of petroleum boiling at higher temperatures than gasoline (e.g., diesel); BTX, benzene, toluene, xylene, compounds that are widely used in the manufacture of organic polymers. Reproduced with permission from A. Sapre, Role of chemical reaction engineering for sustainable growth: One industrial perspective from India, *AIChE J.* **2022**, *69*, e17685.

and methods from earlier chapters of this book and integrates them? Has this chapter helped you to learn to reason from basic principles and methods and see how integrating topics that might at first seem separate is central to developing skills in analysis and design?

PROBLEMS

6.1. Find out who Daniel Fahrenheit was and summarize his design of mercury-in-glass thermometers and their historical significance in medicine and chemical technology.

6.2. Do some research and find out the details about the safety concerns associated with mercury. Find out how the U.S. Environmental Protection Agency has minimized the use of mercury thermometers and how they have been phased out of many industrial applications.

6.3. Find out what a laboratory mercury spill kit is and explain how it works.

6.4. Explain why a mercury thermometer is not used for temperatures of −40°C and lower.

6.5. Design a thermometer like the one considered in Example 6.1, but use ethyl alcohol with a colored dye instead of mercury as the thermometer fluid. Can this be used at temperatures less than −40°C?

6.6. Find out what the alloys constantan, chromel, and alumel are and why they are used in thermocouples.

6.7. Suggest practical issues such as corrosion that may affect thermocouple lifetimes, limiting their long-term applications.

6.8. What are the dimensions of thermal conductivity expressed in terms of mass, length, time, and temperature?

6.9. In the Akita Prefecture in Japan, a traditional seafood soup is cooked at the restaurant table as hot rocks are added to the soup in a wooden container. Assume that a volume of 3 L of the soup in a wooden container is to be heated from room temperature to a few degrees below the boiling point of water. Assume that the rocks are heated in a wood fire and brought to the table. Estimate the mass of rocks needed to heat the soup. State your assumptions.

6.10. Design an experiment to determine the heat capacity of copper, of iron, and of polyethylene.

6.11. Distillation of ethyl alcohol–water mixtures is sometimes carried out with added benzene, which breaks the azeotrope. Do some research to find out what breaking the azeotrope means and how this process works. Further, find out what health/safety concerns are associated with ethyl alcohol that has small amounts of benzene in it resulting from this process.

6.12. Design an ice box for use by campers. What are your design criteria? Consider issues of manufacturing cost, ease of use, proper disposal, and environmental protection.

6.13. Consider a furnace like that described in Example 6.10. If the inside wall temperature is increased to 1,250 K and the thickness of the insulating brick is tripled, but all the other dimensions and the outside wall temperature are unchanged, what are the temperatures between the layers at steady state?

6.14. If the metal rod referred to in Example 6.8 is replaced with aluminum and the dimensions are unchanged and the experiment otherwise conducted in the same way, find the steady-state heat transfer rate through the rod.

6.15.

A. Shipping packs that are plastic-encased and reusable are used in the hot times of the year in packages for overnight transport of live animals and plants. These packs help keep the temperature from becoming too high and endangering the animals and plants; because the packs are temperature controllers, the frozen solid absorbs lots of energy as it turns to liquid at a temperature of about 298 K. When shipments are made in the colder times of the year, the packs are used with the contents initially present as liquid. Explain how these shipping packs work and why the contents are chosen to be biodegradable. Can you find out what material may be contained in the packs? Why are the animals and plants packaged with the shipping packs in boxes that are good insulators?

   B. Find out the details about first-aid applications of such packs.

6.16. A desert water bag made of linen with a volume of 3 L is initially filled with water at 35°C. Estimate how much water would have to be evaporated from it to cool the contents to 25°C.

6.17. Nocturnal cold-blooded animals such as snakes are sometimes found on remote asphalt-paved roads after sundown. Can you figure out why they are attracted to them? And why they are found less frequently on dirt roads and on well-traveled asphalt roads?

6.18. The desert night lizard *Xantusia vigilis* has markedly different light and dark phases. Can you find information illustrating the differences? Can you find information that might suggest a correlation between color and temperature of the animal?

6.19. Do some research and find a small, inexpensive device (a scientific toy) that converts wind energy into light in a small LED. Can you find out where to buy parts to assemble such a device?

6.20. Do a search and find information about calorimetry experiments done with pizza and summarize quantitative conclusions from the reports.

6.21. Look up the properties of asbestos and explain why it is a health hazard and avoided as an insulator today.

6.22. An approximately linear relationship between thermal conductivity and density was reported for 20 different woods at 12 mass% moisture content at an average temperature of 297 K for wood densities in the range of approximately 368–769 kg/m³. The value of the thermal conductivity at the highest wood density was found to be approximately 0.18 watt/(m×K). Determine an equation to relate thermal conductivity to wood density in the range of the data and determine how closely the estimate for Douglas fir at a density of 545 kg/m³ matches the value in Table 6.4. Source of data: Thermal Conductivity of Building Materials, F. B. Rowley and A. B. Algren, Bulletin 12, Engineering Experiment Station, University of Minnesota, 1937.

6.23. The rate of heat transfer by conduction per unit area of a solid depends on the thermal conductivity of the solid, k, the temperature difference ΔT, and the length of the heat conduction path, Δz. Use dimensional analysis to develop an expression for the rate of heat transfer per unit area in terms of the variables stated.

6.24. A hand warmer design includes a supersaturated solution of sodium acetate and a metal disk. Bending the disk creates nucleation sites where the sodium acetate crystallizes. This process gives off heat. If the device is reheated, the sodium acetate dissolves again, and the device can be reused. Find estimates of the needed data and suggest a design that specifies the appropriate dimensions and concentrations and estimate how much energy the device gives off when the crystallization occurs. Explain why supersaturated solutions of sodium acetate can be made.

6.25. Falling ice poses significant hazards in cities. Search for information about the consequences of falling ice and also consider the substantial hazards associated with delivery of blocks of ice to second- and higher-story build-ings, which was important in cities such as New York years ago. Consider the related safety issues as buildings become more energy-efficient.

6.26. Do some research and identify a catalyst for a reaction of formic acid that gives products other than water and carbon monoxide—that is, hydrogen and carbon dioxide. Could the apparatus shown in Example 6.11 be used to determine rates of this reaction? Explain why or why not.

6.27. Consider Figure 6.14 and focus on the black curve, which represents the heat transfer rate. Explain how the heat transfer rate data were used to determine the conversion shown in the gray curve. Make approximate, quantitative statements in your analysis.

6.28. Examine the data of Figure 6.14 and demonstrate the consistency between the differential and integral data. Your calculations will involve substantial error because of the limitations of reading the graph.

6.29. In her Californian novel "Mecca" (Farrar, Straus and Giroux, 2022), Susan Straight wrote about characters trying to keep oranges on trees from freez-ing at 27 degrees, writing that one character sprayed water on some of the trees because the fruit froze at thirty two but the juice would stay sweeter than at dry twenty seven. Explain.

6.30. In her novel "Mecca" (Farrar, Straus and Giroux, 2022), Susan Straight wrote about the use of smudge pots to keep oranges on trees from freezing. Find out what smudge pots are, how they influence heat transfer in orange groves, and why their use is not allowed today in California.

6.31. Compare Fourier's law of heat conduction with Ohm's law.
   A. What are the driving forces for conduction of heat, on the one hand, and of electrical current, on the other? Can you define a resistance to heat conduction in relation to thermal conductivity that is analogous to an electrical resistance?
   B. Thermal conductivities are temperature dependent; does a comparable statement pertain to the resistance of an electrical conductor?

6.32. Using the data characterizing the temperature dependence of cricket chirp-ing presented in Chapter 5, find an approximate equation that relates the temperature to the chirping rate.

6.33. Attempts to control nuclear fusion and develop practical, economical sources of energy from it have led to recent news reports that experiments have been done whereby more energy was produced in a nuclear fusion reaction that was introduced into the experimental reactor. Thus, energy balance data have been in the news. Do some research and find out what the ratio of energy emitted to energy introduced was and how energy was introduced into the reactor.

6.34. Can you do some research and find order-of-magnitude estimates of some dimensionless numbers that provide some perspective on energy availabil-ity and use by humankind:
   A. The ratio of the rate of energy absorption by the earth to the rate of energy release by the sun.
   B. The ratio of reflection of sunlight energy by the earth to sunlight energy incident on the earth.

C. The ratio of incident energy from the sun collected by solar cells to the energy absorbed by the earth.

D. The efficiency of a modern solar cell that collects energy from the sun.

E. The efficiency of a leaf as a solar energy collector.

6.35. Can you identify some dimensionless groups that include thermal conductivity as a term and that are important in engineering analysis and correlations? Considering the dimensions of thermal conductivity, can you suggest some other physical properties that might appear in dimensionless groups that include thermal conductivity?

6.36. Two important dimensionless groups used by chemical engineers are the Prandtl number and the Nusselt number. Find out what these are, show that they are dimensionless, and explain briefly why they are important in correlations involving heat transfer. The explanation will require some research to define a new term.

6.37. Do some research and find a dimensionless correlation that involves the Prandtl number and one that involves the Nusselt number. What other familiar dimensionless groups appear in these correlations? Do you have an example that accounts for heat transfer to a flowing fluid? Why does it make sense that such a correlation involves the dimensionless groups that you see in the correlations? Can you find a correlation that involves all of the following: the Reynolds, Prandtl, and Nusselt numbers?

6.38. In this chapter, heat of mixing was illustrated for potassium hydroxide and water. Another good example would involve mixing of an aqueous solution of sulfuric acid and water and checking the prediction with a calorimeter. However, suggesting such a prediction and a corresponding calorimeter experiment involving pure sulfuric acid and water does not make good chemical sense. Check the chemistry of sulfuric acid and explain why not.

6.39. Find out what a shell and tube heat exchanger is; sketch it, and explain how it works. What dimensionless groups do you expect to be significant in correlations predicting the heat exchanger performance?

6.40. Suppose that, in the experiment described in Example 6.13, the experimenter had mistakenly used 3.12 g of sodium chloride instead of the intended 3.12 g of glucose. What would have been observed in the experiment?

6.41. Can you figure out why the heat of vaporization of water is so much greater than that of methyl alcohol and why the heat of vaporization of methyl alcohol is greater than that of ethyl alcohol? (This is a chemistry question.) Can you make an educated guess about the heat of vaporization of ethane and check it against reported values?

6.42. Considering some history of the Inuit, design an igloo for Arctic hunters made with blocks of compacted snow. What advantages does compacted snow have over ice? What safety features need to be incorporated in your design, especially if a heater is to be used in the igloo?

6.43. A phase change we have not considered is sublimation. Find out what that term means, and determine the heat of sublimation of ice.

6.44. Check recent sources providing information about "smart" windows that provide temperature control of buildings, and find out what they are made of and how they work.

6.45. Find out what is meant by the term "circular economy" and how such an economy would benefit the environment and humankind.

6.46. In the winter, Monarch butterflies congregate by the thousands and cloak—they organize themselves into layers. Can you explain why?

Index